Physics: More Than You Know

Sandeep Chavan

Published by Gyrus Vision, 2026.

PHYSICS: MORE THAN YOU KNOW

First edition. June 21, 2026.

ISBN: 978-8168977631

Written by Sandeep Chavan.

Table of Contents

To every student who found the correct answerand still had the courage to ask, "But what is actually happening?"

"The deepest task of physics is not merely to measure reality, but to understand how reality becomes measurable."

— **Sandeep Chavan**

Opening Note

Physics Before It Became a Formula

I have taught physics for many years, and one thing has always remained clear to me: most students encounter physics only after it has already been converted into definitions, quantities, symbols, equations, diagrams, and numerical problems. They learn that force is measured in newtons, energy in joules, pressure in pascals, velocity in metres per second, and temperature in degrees. They learn formulas, substitute values, rearrange terms, draw free-body diagrams, and calculate the required answer. All of this is necessary. Without measurement, physics would lack precision. Without mathematics, it would struggle to express relationships clearly. Without equations, much of modern engineering, technology, astronomy, medicine, and communication would be impossible.

Yet something important often disappears during this process. Before force became a quantity, it was an experience. Before pressure became a formula, it was resistance. Before temperature became a number, it was warmth and cold. Before motion was divided into displacement, velocity, and acceleration, it was simply change unfolding before awareness. Human beings did not first encounter physics in laboratories or textbooks. We encountered it through the body and through our interaction with the surroundings. We lifted stones, pushed objects, pulled ropes, slipped on wet ground, felt the heat of fire, watched water flow, heard thunder, observed shadows, and looked at the night sky. Long before humanity created scientific vocabulary, it was already living inside physical consequences.

Science did not create those consequences. It observed them, organized them, named them, compared them, measured them, and eventually expressed their regularities through mathematics. That achievement is extraordinary. The movement from experience to measurement transformed human civilization. It allowed knowledge to travel beyond memory and intuition. It made prediction possible. It enabled one person's observation to be verified by another. It allowed humanity to build machines, generate electricity, communicate across continents, examine atoms, launch satellites, and explore regions of the universe that no human body could directly reach.

But the success of scientific vocabulary also created a subtle confusion. Over time, we began treating the quantities used to describe reality as though they were reality itself. We speak of force as if force were a separate substance acting upon an object. We speak of energy as though it were a material stored inside things. We speak of fields as if naming a field automatically explains what exists. We speak of spacetime as though a successful geometrical description settles every question about the structure of reality. The language becomes so familiar that we stop noticing the difference between the phenomenon and the description.

A quantity can be measured. An equation can express a relationship. A model can reproduce observed behaviour. A theory can predict an outcome. But none of these automatically tells us what reality fundamentally is. This is not a criticism of physics. It is an attempt to appreciate physics more deeply. Newtonian mechanics was not made useless when Einstein introduced a broader description of gravity. Classical physics did not disappear when quantum theory entered the picture. Thermodynamics remained powerful even after molecular theory gave deeper meaning to heat and temperature. Science often advances not by declaring earlier descriptions worthless, but by discovering the larger structures within which those descriptions remain valid.

Force is a good example. Force began as a natural human experience of push, pull, effort, resistance, weight, tension, and impact. Classical mechanics transformed that experience into a precise and highly useful quantitative relationship. Later developments in physics began treating many apparent forces through fields, potentials, geometry, and interactions. Gravity, experienced in ordinary life as a downward pull, is described in general relativity through spacetime geometry. Electrical and magnetic effects are described through electromagnetic fields. At deeper levels, modern physics increasingly speaks through symmetries, states, probabilities, interactions, and field configurations. The phenomenon remains, but the vocabulary changes. The description moves deeper.

This raises an important question: what if physics is more than its quantities? What if force, pressure, charge, momentum, temperature, and entropy are not the basic furniture of reality, but carefully developed variables used to describe the ways reality manifests? What if equations are not the final architecture of nature, but highly successful maps of recurring consequences? And what if physics, at its deepest level, is not merely the science of measurement, but the study of how existence organizes, interacts, transforms, and becomes observable?

These questions do not require us to abandon mathematics. On the contrary, mathematics becomes even more valuable when we understand its proper role. It is among the most powerful languages humanity has created. But a language should not be confused with the reality it describes. A map is not useless because it is not the territory. Its usefulness depends precisely on how accurately it represents the territory. Likewise, physical quantities are not diminished when we call them descriptive. Their value lies in their ability to capture stable relationships within nature. But the description still belongs to a different layer from the underlying process.

This book begins from that distinction. Reality comes first. Then come interaction, change, pattern, and consequence. After that come observation and measurement. Only then do quantities, equations, and theories emerge. The difficulty begins when this order is reversed—when a mathematical symbol is treated as more real than the process it was introduced to describe.

Physics has always needed imagination. Every major theory required someone to look beyond what was immediately visible. Newton had to imagine universal relations extending beyond ordinary experience. Faraday had to think in terms of fields. Einstein had to rethink space, time, motion, and gravity. Quantum theory demanded conceptual structures that challenged everyday intuition. Imagination is therefore not the enemy of science. But imagination must retain its scientific position. Observation, measurement, mathematical modelling, theoretical inference, philosophical interpretation, and speculation are not the same layer of knowledge.

An event may be observed directly. A quantity may be introduced to measure one aspect of it. An equation may successfully express the regularity. A theory may organize several such regularities into one framework. But the story we attach to the theory may still remain interpretive. A mathematical possibility may be elegant without being physically confirmed. A model may predict accurately without revealing what fundamentally exists. A theory may work within its domain while remaining incomplete as an account of reality.

Physics does not lose its scientific character when it asks philosophical questions. Philosophy has always accompanied serious physics. It loses clarity when philosophical interpretations are presented as measured facts. When those boundaries are ignored, physics can move quietly from science into speculation, from speculation into metaphysics, and from metaphysics into mythology or science fiction—while continuing to speak in the authoritative language of equations.

The problem is not speculation itself. Speculation often opens new doors. The problem is unmarked speculation: presenting what is imagined, inferred, or interpreted as though it had already been observed. A disciplined reader of physics should therefore keep asking: what was actually observed, what was measured, what quantity was introduced, what does the equation successfully predict, and what interpretation has been added afterward. These questions do not weaken science. They protect it. They also protect curiosity. When a scientific explanation is treated as final merely because it is mathematically successful, inquiry stops too early. But when the limits of a theory are acknowledged, space remains open for deeper investigation.

This book does not claim that modern physics has failed.

Modern physics has succeeded beyond anything earlier civilizations could have imagined. It does not argue that equations should be replaced by vague philosophy. It does not dismiss Newton, Maxwell, Einstein, thermodynamics, quantum mechanics, or field theory. Instead, it asks whether the scope of physics may be broader than the form in which it is commonly taught. Perhaps physics is not only about how much. Perhaps it is also about how something becomes measurable. Perhaps it is not only about quantities. Perhaps it is about manifestations.

Matter does not carry labels. A stone does not know mass, momentum, force, or gravitational potential. A planet does not know that it is following an orbit. An electron does not read the equation used to predict its behaviour. Things respond to surrounding conditions, differences, gradients, configurations, and interactions. We observe those responses, identify regularities, and express them through physical quantities. From this perspective, force does not need to exist as a separate object travelling from one body to another. It can be understood as a description of how motion changes under particular conditions. Temperature is not an independent substance residing inside matter. It represents a measurable manifestation of the energetic state of a system. Pressure describes the distributed consequence of interactions. Momentum captures a relationship between material form and motion. Entropy describes changes in configuration, distribution, and accessibility. The quantities are real as measurable relationships. But measurable reality and fundamental reality may not be identical.

This opens a larger question. If physical quantities describe manifestations, what produces the manifestations? If fields explain forces, what gives rise to fields? If matter is a structured form of energy, what allows that structure to emerge? If spacetime describes relations among events, is it the final foundation—or another powerful descriptive layer? These are not entirely new questions. Science has repeatedly encountered them in different forms. Every successful theory resolves one layer while revealing another. The deeper physics looks, the less reality resembles the collection of isolated solid objects suggested by ordinary experience.

This book will gradually move toward the possibility that matter, fields, interactions, and physical quantities may arise through a deeper universal substrate. In Universal Energy Dynamics, I refer to this foundational reality as Structured Vacuum Energy, or SVE. But this is not intended to be a technical book on UED. The purpose is more modest. It is to reopen the reader's relationship with physics. It is to move behind textbook definitions and ask what they actually represent. It is to distinguish reality from the vocabulary developed to describe it. It is to show that questioning a concept does not mean rejecting its usefulness. Sometimes the deepest respect we can offer science is to refuse to stop at the first successful description.

Physics began in wonder before it became calculation. A child watches an object fall and asks why. A student learns the equation and is taught how. A scientist searches for the structure beneath the equation. A philosopher asks what the explanation truly means. These are not competing activities. They are different layers of the same human inquiry. The problem begins only when one layer declares itself complete.

This book is therefore an invitation—not to distrust physics, but to see more within it. Not to move away from quantities, but to understand where they stand. Not to abandon equations, but to look through them. Not to reject imagination, but to distinguish imagination from evidence. Physics may be one of humanity's most precise sciences. But precision does not reduce its depth. The more carefully we look, the more we may discover that physics is not merely a collection of quantities describing the universe. It is humanity's evolving attempt to understand how reality becomes form, interaction, pattern, measurement, experience, and consequence. And that may be far more than we currently know.

Preface

This book began in the classroom, not in a laboratory or conference hall.

For more than fifteen years, I have taught physics to students preparing for board examinations, engineering entrance tests, medical entrance tests, and future careers in science and technology. During that time, I have watched students become increasingly skilled at solving numerical problems. They learn formulas, units, diagrams, derivations, and standard methods. With sufficient practice, many can calculate quickly and accurately.

Yet I have also seen something gradually disappear: **the physical event itself.**

A student may know how to calculate force but may never seriously ask what force is. A student may use the equation for pressure without seeing the collective interaction hidden behind the value. A student may learn that temperature is measured in kelvin and still imagine heat as a substance stored inside matter. A student may solve problems involving electric and magnetic fields without asking what kind of physical reality a field represents.

This is not the fault of the student. Modern science education must cover large syllabi within limited time. Examinations reward correct methods and accurate answers. Teachers must prepare students for competitive systems. Textbooks condense centuries of scientific struggle into finished definitions and formulas.

All of this is understandable. But when physics is introduced only through its completed vocabulary, students may learn how to use the language without ever encountering the questions that created it. They meet the equation after the mystery has been removed.

This book is an attempt to restore some of that mystery—not by making physics vague, but by examining it more carefully.

By *mystery*, I mean the honest boundary between what has been successfully described and what has not yet been fully understood. Physics has achieved extraordinary precision. It predicts planetary motion, controls electrical systems, supports modern communication, explains material behaviour, guides medical technology, and examines regions of the universe far beyond direct human reach. Its success is beyond question.

But success in prediction does not automatically settle every question of meaning. A theory may work and still leave its ontology open. A quantity may be measurable and still not be fundamental. An equation may describe a relation without revealing the entire physical process beneath it. A field may organize observed behaviour while the deeper nature of the field remains uncertain.

This distinction is central to the book.

I have not written these chapters to argue against established physics. I continue to teach Newtonian mechanics because it works. I teach energy, momentum, pressure, gravitation, thermodynamics, electricity, magnetism, optics, and modern physics because these frameworks are essential. A student cannot become a serious thinker by rejecting knowledge that has already survived observation and experiment. But neither should serious thinking stop at the first successful description.

Physics has repeatedly advanced by re-examining concepts that once appeared complete. Force gave way to field-based explanations. Absolute space and time gave way to relativity. Solid matter gave way to atoms, particles, and quantum fields. The vacuum ceased to look like simple emptiness. The history of physics shows that a concept can remain useful even after its fundamental status changes.

That possibility is one of the reasons this book exists.

The central question is not whether physical quantities are false. They are not. The question is where they stand. Mass, force, energy, charge, temperature, pressure, momentum, and entropy are part of the highly successful vocabulary of physics. They describe stable and measurable aspects of physical behaviour. But are they the deepest contents of reality? Or are they the variables through which observers describe recurring manifestations of a deeper physical structure?

This book develops the second possibility. It treats physical quantities as the *manifestation vocabulary* of science.

That phrase should be read carefully. It does not mean that quantities are subjective inventions detached from nature. Reality constrains every reliable measurement. A bridge will not remain standing because an engineer prefers a different answer. A spacecraft will not reach its destination because a calculation is philosophically attractive. Nature decides whether the relation is correct. But the units, symbols, definitions, and conceptual categories remain human constructions developed in response to objective regularities.

The phenomenon is not invented. The vocabulary is.

This distinction allows us to respect scientific knowledge without confusing representation with reality.

The book begins with force because force is one of the most familiar physical ideas. We experience push, pull, impact, strain, weight, and resistance through the body. The concept therefore appears almost self-evident. But the closer physics examines interaction, the less force resembles a separate invisible substance. It may instead describe changes arising through fields, gradients, constraints, deformation, momentum exchange, or geometry.

From force, the argument moves toward nature and scientific language, then toward physical quantities, fields, mathematics, manifestation, and the question of deeper physical structure. The later chapters introduce **Universal Energy Dynamics (UED)** and **Structured Vacuum Energy (SVE)**.

I want to state clearly how these ideas should be approached. UED is a developing foundational framework. SVE is a proposed physical foundation. Neither should be accepted merely because the ideas offer conceptual unity. A broad explanation is not automatically a scientific theory. A scientific framework must define its terms, develop formal relations, recover established results, produce distinctive predictions, and remain open to experimental failure.

I have introduced UED in this book because the earlier chapters naturally raise the question of what may lie beneath matter, fields, and physical quantities. SVE is one possible answer. It is not presented as the only possible answer, nor as a completed replacement for modern physics. The scientific journey from conceptual possibility to established theory is demanding. It should be. A framework that seeks to describe reality must allow reality to correct it.

This book is therefore written for several kinds of readers:

- Students who have learned formulas but continue to ask what they mean.
- Teachers who want to restore conceptual depth to the classroom.
- Engineers who use physical quantities every day and occasionally wonder what lies beneath their operational definitions.
- Science lovers who respect modern physics but do not believe that every foundational question has been closed.
- Readers curious about UED but needing a clear doorway before entering its broader and more technical ideas.

No advanced mathematics is required. The book is conceptual rather than computational. But it should not be mistaken for a rejection of mathematics. Mathematics is indispensable to physics. The argument here is only that mathematical representation and physical reality occupy different layers. An equation can be both extraordinarily accurate and ontologically incomplete. That possibility should not frighten science. It should keep science alive.

The reader may disagree with some interpretations in the chapters ahead. That is acceptable. The purpose is not to secure agreement at every step. The purpose is to encourage a more disciplined kind of questioning:

- What was actually observed?
- What was measured?
- What quantity was introduced?
- What does the equation predict?
- What interpretation has been added?
- What remains unresolved?

These questions are useful not only for UED, but for every scientific framework. They protect us from two opposite errors: blind rejection—the belief that because a theory is incomplete, it must be useless; and blind finality—the belief that because a theory works, it must describe the deepest reality exactly as it is.

Science grows between these extremes. It trusts evidence without worshipping vocabulary. It uses models without confusing them with the whole. It allows imagination without presenting imagination as observation. It remains open without becoming careless.

I have deliberately written this book in accessible language. Some ideas recur from different directions. This is intentional. Foundational questions are rarely understood through one definition alone. They become clearer when viewed through experience, history, quantity, theory, language, and consequence.

The reader need not agree immediately. It is enough to pause.

Look at the stone before calculating its fall. Look at the hand before naming the force. Look at the material before assigning its temperature. Look at the event before reaching for the equation.

Then use the equation. Measure carefully. Calculate honestly. Respect the result. But after the answer has been obtained, return once more to the event.

Ask what the quantity has captured. Ask what it has left out. Ask what physical structure makes the measurable result possible.

That return—from formula to reality—is the real journey of this book.

Physics may be among humanity's most precise achievements. But precision does not make it shallow. The deeper we look, the more we discover that every quantity opens into a larger question, every theory stands within a wider reality, and every correct answer may still contain the beginning of another inquiry.

That is why physics remains more than we know.

— Sandeep Chavan

How to Read This Book

This book is not a textbook in the conventional sense. It does not teach physics chapter by chapter through definitions, derivations, and numerical exercises. It assumes that many readers already possess some familiarity with terms such as force, energy, mass, temperature, pressure, field, gravity, and motion.

Its purpose is to return to those familiar terms and ask what they actually mean.

The book can be read from beginning to end, and that is the recommended approach. Its argument develops gradually. The early chapters begin with experience and scientific vocabulary. The middle chapters move toward fields, force, and mathematics. The later chapters introduce manifestation, deeper structure, Structured Vacuum Energy, and the possible future of physics. Each stage prepares the next.

Readers who move directly to the chapters on UED and SVE may understand the definitions, but they may miss why those ideas arise. The deeper proposal becomes meaningful only after the distinction between reality, measurement, quantity, equation, theory, and interpretation has been established.

Read Conceptually, Not Competitively

There are no numerical problems to solve. You are not required to memorize definitions. The aim is not to finish quickly or agree with every sentence.

Pause whenever a familiar scientific word begins to feel unfamiliar. Ask yourself:

- What do I normally imagine when I hear this term?
- Was that image measured, or was it taught through analogy?
- Does the quantity describe an object, a relation, a response, or a collective condition?
- What does the equation successfully predict?
- What remains open after the calculation?

These pauses are part of the reading.

Keep the Levels Separate

Throughout the book, several levels of scientific knowledge appear repeatedly:

- physical event
- observation
- measurement
- quantity
- equation
- theory
- interpretation
- speculation

Try not to combine them too quickly.

An event may be observed directly. A quantity may be introduced to measure one aspect of it. An equation may express a reliable relation. A theory may organize several equations and observations. An interpretation may then propose what the theory means about reality.

These levels are connected, but they are not identical. One of the main purposes of the book is to help the reader notice when a description has quietly been treated as the thing described.

Do Not Read It as an Attack on Physics

The book questions the finality of certain concepts, not their usefulness.

Force remains necessary. Energy remains measurable. Fields remain central. Mathematics remains indispensable.

Relativity, thermodynamics, quantum theory, and classical mechanics retain their scientific authority within their established domains. The argument is not that modern physics is false. The argument is that successful descriptions may still stand within a deeper reality.

Read every criticism in that spirit. Questioning whether force is fundamental does not make force calculations invalid. Asking what a field is does not deny field theory. Distinguishing mathematics from reality does not weaken mathematics. It clarifies the role of each.

Treat Analogies Carefully

The book uses familiar examples such as maps, landscapes, waves, flames, whirlpools, tables, stones, and musical scores. These analogies are intended to help the reader see a conceptual relation. They should not be treated as literal physical models.

When matter is compared with a stable pattern in a whirlpool, the claim is not that matter is made of water. When a theory is compared with a map, the claim is not that scientific theories are merely drawings.

An analogy reveals one aspect while hiding others. Use it as a bridge, then leave it behind.

Notice the Difference Between "Real" and "Fundamental"

A recurring idea in this book is that something can be real without being fundamental.

Temperature is real. Pressure is real. A solid object is real. A force measurement is real.

But each may arise from deeper physical organization. Calling something emergent does not make it imaginary. It means that the property appears at a particular level because of relations or dynamics at another level.

This distinction is important throughout the book. The question is often not whether something exists. The question is how it exists and at what level.

Read the Interludes as Pauses

The two interludes are not separate chapters containing new arguments. They are transitions.

When Quantities Begin to Disappear marks the movement from familiar measurements toward fields and distributed structure. *After the Equation* marks the movement from mathematical representation toward manifestation science.

Read them slowly. They are meant to gather what has already been established and prepare the reader for the next conceptual shift.

Approach UED with Openness and Discipline

The later chapters introduce Universal Energy Dynamics and Structured Vacuum Energy. These ideas should neither be dismissed automatically nor accepted prematurely. Read them as a developing foundational proposal.

Ask the same questions that the book applies to every theory:

- What is being proposed?
- Which established observations motivate it?
- What does it explain differently?
- Which existing results must it preserve?
- What mathematical structure is still required?
- What prediction could distinguish it?
- What evidence could challenge it?

The purpose of introducing UED is not to demand belief. It is to offer one possible direction after the limitations of force-centred, object-centred, and purely quantitative descriptions have been examined.

The reader is free to accept, reject, modify, or suspend judgment.

Return to the Book After Reading It

Some chapters may appear simple on the first reading because they begin with ordinary experiences:

- A hand presses a wall.
- A stone falls.
- A thermometer gives a reading.
- A magnet affects iron.

But the questions beneath these examples are foundational.

After completing the book, return to one familiar concept—force, mass, temperature, pressure, field, vacuum, or energy—and read the relevant chapter again. The second reading may feel different because the later chapters change the meaning of the earlier ones.

The book is designed to move in both directions. The early chapters prepare the deeper proposal. The deeper proposal then reopens the early chapters.

Keep One Question Alive

You do not need to retain every argument. You do not need to remember every historical example. You do not need to adopt every interpretation.

Keep one question alive:

What is reality doing before we name, measure, and calculate it?

Use the quantity. Work with the equation. Respect the evidence. But continue looking through the vocabulary toward the physical event.

That is the most useful way to read this book. It is also one of the most useful ways to read physics.

Before the Formula

"Physics does not begin with the equation. It begins with the event that made the equation necessary."

"Nature produces consequences; human intelligence converts them into quantities."

"A formula may complete the calculation while leaving the physical question entirely open."

"The universe does not obey our laws. Our laws survive only when they remain faithful to the universe."

"The deepest task of physics is not merely to measure reality, but to understand how reality becomes measurable."

"Every quantity is a window into reality—but no window should be mistaken for the whole sky."

— ***Sandeep Chavan***

Chapter 1 — Did We Discover Force or Invent It?

Before the Word Existed

Place your palm against a wall and push. The wall does not move, yet something unmistakable happens. Your muscles tighten, your arm becomes rigid, pressure develops in the hand, and the body registers effort even though no visible displacement occurs. Now push an empty chair across the floor. The chair moves, the resistance is smaller, and your body adjusts almost immediately. Push a loaded cupboard, and the experience changes again. The same human action—pushing—produces different consequences depending on the object, the surface, the direction, and the surrounding conditions.

Long before anyone wrote an equation, human beings already knew these differences. They knew that a stone was harder to move than a dry leaf. They knew that carrying water uphill required more effort than walking downhill. They knew that a stretched rope pulled back, that falling objects struck the ground, that wind bent trees, and that animals resisted when dragged. No one needed the word *force* to experience any of this. The experience came first. The word came later.

This simple order matters because modern students usually encounter it in reverse. They first meet force as a definition: a push or pull capable of changing the state of motion of an object. Soon afterward, force becomes a symbol, usually F, followed by units, diagrams, laws, and numerical problems. By the time the student learns to calculate force, the original experience has already been compressed into a quantity. That compression is useful. It makes comparison possible. It allows the effect of a moving truck to be distinguished from that of a rolling ball. It allows engineers to design bridges, machines, brakes, aircraft, and buildings. It gives physical experience a precise and transferable language. But it can also hide an important question: what exactly did we discover? Did we discover a thing in nature called force, or did we invent a concept that allows us to describe recurring changes in matter?

The difference may appear philosophical, but it reaches directly into the foundations of physics. If force exists independently as a physical entity, then physics must eventually explain what force is, where it comes from, and how it travels between objects. But if force is a descriptive quantity, then the deeper scientific task is different. We must explain the conditions, gradients, fields, and interactions that produce the response we describe as force. The falling stone does not announce that gravity is acting upon it. The stretched spring does not calculate tension. The planet does not know it is being pulled toward the Sun. Nature behaves. We observe. Science names the pattern. The question is whether, after naming the pattern, we mistake the name for the cause.

1. Experience Came Before Measurement

The earliest physics was not written in symbols. It was carried in the body. A person lifting a stone learned something about weight. A hunter throwing a spear learned something about direction, speed, balance, and resistance. A farmer pulling a plough learned that wet soil and dry soil responded differently. A sailor learned that wind could move a vessel, waves could redirect it, and water could resist its motion. These people may not have possessed a formal theory of mechanics, but they were not physically ignorant. Their knowledge was embedded in action.

The body continuously detects differences in the surrounding world. It senses when more effort is required, when balance is disturbed, when motion begins, when motion stops, and when contact becomes painful. Muscles, joints, skin, vision, and the inner ear together produce a direct awareness of physical interaction. This is why force feels so natural to us. It emerges from conscious experience. When we push something, we feel resistance. When we carry a load, we feel strain. When we jump, we feel the ground beneath us. When a moving object strikes us, we experience impact. The concept of force appears almost unavoidable because our body interprets the world through interaction.

But the object on the other side of the interaction does not experience it in the same conscious sense. A cupboard does not feel heavy. A stone does not feel that it is falling. A metal rod does not experience tension as discomfort. These objects respond according to their structure and surrounding conditions. Their internal arrangements change, their motion changes, their shape may deform, and energy may redistribute. But the human language of push, pull, strain, impact, and force comes from the way conscious beings encounter these responses.

This gives force two distinct histories. The first is biological and experiential. The second is scientific and quantitative. The biological history is ancient. It begins wherever living organisms interact with resistance. Even animals understand, in practical terms, that some objects are easier to move than others. A bird adjusts its wings against the air. A predator judges how strongly it must leap. An insect navigates surfaces, currents, and obstacles. The scientific history begins when humans attempt to separate the experience from the individual body and express it in a form others can compare. "My effort was greater" is personal. "This object accelerated by a certain amount under specified conditions" is measurable. That transition is one of the great achievements of science. It removes ambiguity. It transforms private sensation into public knowledge.

Yet something is lost during every transformation of experience into measurement. The quantity cannot contain the entire experience from which it emerged. A reading on a scale does not contain the strain in the arm. A force value does not contain fear, effort, pain, or intention. It captures only the measurable physical relation selected for study. This is not a weakness. It is how science achieves precision. The problem begins only when the selected measurement is treated as the whole reality.

Force, then, did not begin as a hidden substance waiting to be found. It began as a recurring pattern within interaction. Human beings encountered that pattern physically, then isolated one aspect of it, named it, and developed methods to measure it. That does not make force imaginary. A currency is invented, yet economic consequences are real. A map is constructed, yet the landscape exists. A unit is defined, yet the difference it measures can be objective. Force may be real in this relational sense: not necessarily as a separate thing, but as a reliable description of how physical systems change. Before asking what force is, therefore, we should first ask what kind of reality a scientific quantity possesses.

2. Push, Pull, Resistance, and the Conscious Body

Textbooks often define force as a push or a pull. The definition is helpful because it connects an abstract quantity to familiar experience. A child already understands pushing and pulling long before learning mechanics. But the simplicity of the definition also conceals a deeper problem. Push and pull are human descriptions of interaction.

Suppose you pull a suitcase using a handle. From your perspective, your hand pulls the handle, the handle pulls the suitcase, and the suitcase moves. The event appears to contain a sequence of actions passing from one object to another. Yet when physics examines the interaction more closely, the ordinary language becomes less straightforward. Your fingers press against the material of the handle, the material deforms slightly, internal electromagnetic interactions resist further overlap, and the structure transmits changes through the handle into the suitcase. The suitcase responds according to its mass, the friction at the wheels, the angle of the pull, and the condition of the surface. What ordinary language compresses into one word—pull—is actually a coordinated chain of material responses.

The same applies when we press a hand against a table. At the level of everyday experience, the table simply stops the hand. At a deeper level, the atoms of the hand do not pass through the atoms of the table because of the structure of matter and electromagnetic interactions. What we experience as solidity is not merely two solid substances touching in the naïve sense. It is a regulated response arising from the organization of matter. Our conscious experience, however, receives the entire process as resistance. The brain does not present a quantum-field account to awareness. It presents pressure, hardness, discomfort, or support. Biological systems are designed to respond rapidly, not to reveal the complete physical architecture of every interaction.

This distinction helps explain why force became such a central idea in classical physics. It matches human experience extremely well. We feel that we cause motion by exerting effort. We feel that objects oppose us. We feel that gravity pulls us downward. We feel that a moving vehicle throws us sideways when it turns. The force vocabulary organizes these experiences into a coherent mechanical system. But conscious experience can also mislead. When a bus suddenly accelerates, passengers feel as though they are pushed backward. No external agent may actually be pushing them toward the rear. Their bodies tend to preserve their previous state of motion while the bus moves forward beneath them. The sensation is real, but the immediate interpretation may be incomplete.

Similarly, a person standing on Earth feels stationary, although Earth rotates and moves around the Sun. The body reports local stability, not the full cosmic motion. Physics repeatedly advances by respecting experience while refusing to be imprisoned by it. The body is the first instrument, but it is not the final instrument. This creates an important distinction between experienced force and formulated force. Experienced force belongs to conscious interaction. It includes strain, effort, pressure, impact, and resistance. Formulated force belongs to scientific description. It relates measurable changes in motion to the physical properties and conditions of a system. The two are connected, but they are not identical.

A person can experience great muscular effort while pushing a wall that does not move. In ordinary speech, the person is clearly exerting force. Yet no mechanical work is done on the wall if there is no displacement. The body consumes energy, but the external object may remain nearly unchanged. This small example already reveals that our everyday categories do not map perfectly onto scientific ones. Effort is not the same as work. Pressure is not the same as force. Weight is not the same as mass. Resistance is not always friction. Scientific language separates what experience tends to combine. That separation is useful, but it should not make us forget the origin of the concept. Force entered human understanding through conscious interaction before it entered mechanics as a quantity.

Perhaps this is why force feels more ontologically real than some other scientific terms. We do not merely calculate it. We seem to feel it. But what we feel may be the body's interpretation of energy transfer, structural resistance, acceleration, deformation, or changing conditions—not an independent substance called force.

3. From Motion to a Mathematical Quantity

The scientific formulation of force did not appear fully formed. For centuries, thinkers struggled to understand why objects move, why they stop, and whether continued motion requires a continued cause. Ordinary experience appeared to show that moving objects naturally came to rest. Push a cart and stop pushing; eventually the cart stops. Throw a stone; it falls. Slide a block; friction slows it. It was therefore tempting to conclude that motion itself required a sustaining force.

The difficulty was that ordinary experience always included hidden conditions. Friction, air resistance, surface roughness, and deformation were constantly present. These effects made it difficult to imagine motion without resistance. Galileo's work helped change the question. Instead of treating rest as the natural condition of matter, he examined how motion might continue when resistance was reduced. Through idealized reasoning and experiments involving inclined planes, he moved physics toward the idea that an object could preserve its motion unless something altered it. This was a profound conceptual shift. Force no longer needed to explain motion itself. It needed to explain a change in motion.

Newton gave this transition its classical form. In his mechanics, force became connected to changes in momentum and, in familiar simplified situations, to mass and acceleration. The equation did not merely define a number. It reorganized the meaning of cause in mechanics. A moving object did not require a continuous force simply to keep moving uniformly. A force was required when its motion changed. This distinction separated force from muscular intuition. In everyday life, maintaining motion often requires effort because resistance is present. In ideal mechanics, uniform motion continues without a net external force. Physics had moved beyond immediate experience by analysing the conditions hidden within experience.

Yet even Newton's success did not settle what force fundamentally was. His equations provided extraordinary predictive power. They described falling bodies, projectiles, planetary motion, tides, and mechanical systems within a unified framework. The same mathematical structure could be applied on Earth and in the heavens. This was one of the great achievements in the history of science. But predictive unity is not the same as ontological explanation. Newton could express how gravitational interaction depended on masses and distance. He could calculate trajectories with remarkable accuracy. What remained difficult was explaining how one body influenced another across apparently empty space. The mathematics worked. The mechanism remained obscure.

This distinction is crucial for the argument of this book. A scientific concept can be operationally successful before its deeper meaning is settled. Force became measurable through its effects. Scientists could infer it from acceleration, deformation, balance, or momentum change. Engineers could use it with confidence. Yet the quantity's usefulness did not automatically prove that force existed as an independent physical substance. In fact, Newtonian mechanics treats force more like a relational input within a law than like a material object. Force is identified through what it does within the mathematical framework.

This creates a subtle circularity if handled carelessly. We say force causes acceleration. Then we calculate force by measuring acceleration. The relation is immensely useful, but the equation alone does not tell us whether force is a separate entity, an interaction, a field effect, a gradient, or simply a descriptive variable connecting observed changes. The equation establishes a structure of relation. Interpretation comes afterward.

This is where textbooks often compress history too aggressively. Students are shown the final formula but not the conceptual struggle that made it necessary. They may come to believe that force was simply discovered in nature in the same way one discovers a new mineral. But force was not found lying inside an object. It was formulated to organize observed changes. This does not reduce its scientific reality. On the contrary, it shows the sophistication of scientific construction. Science often advances by creating concepts precise enough to capture patterns that ordinary language cannot manage. Velocity, acceleration, momentum, field, potential, and entropy are all examples of this creative discipline. Science discovers regularities. It invents the vocabulary used to express them. Force belongs to both processes. The behavioural pattern was discovered. The conceptual quantity was constructed.

Confusing these two achievements leads us to ask the wrong question. Instead of asking where force as an object came from, we should ask what underlying structure produces the changes that the force concept successfully measures.

4. Newton's Dilemma: Prediction Without Mechanism

The success of universal gravitation created one of the most revealing dilemmas in physics. Newton's theory described the gravitational relation between bodies with extraordinary precision. The same framework that explained a falling apple could also describe the motion of the Moon and planets. This was revolutionary because it dissolved the old separation between earthly motion and celestial motion. Nature appeared unified.

But the unity came with a conceptual problem. How could the Sun influence Earth across an enormous distance? How could Earth influence the Moon through empty space? What carried the interaction? The equation specified how gravitational influence varied with mass and distance, but it did not provide a mechanical picture of what travelled between the bodies. To many thinkers, action across empty space seemed deeply troubling.

Newton himself was cautious about pretending to know more than the mathematics and observations justified. He developed a powerful law without claiming to possess a final account of gravity's underlying cause. That caution is scientifically important. He did not allow the success of the description to erase the incompleteness of the explanation. Later generations sometimes treated Newtonian force more concretely than Newton himself had. Once a theory becomes standard, its conceptual uncertainty tends to disappear from teaching.

Students learn to calculate gravitational force and may assume that the force itself has been fully explained. But an equation that predicts a relation does not necessarily reveal the architecture producing it.

This became clearer as physics developed the concept of the field. Instead of imagining one body reaching mysteriously across space to pull another, field language placed physical structure throughout the surrounding region. A mass, charge, or magnet was associated with a field, and another object responded according to the local field conditions at its position. This was a conceptual improvement because the interaction was no longer described only as a direct relation between distant objects. The space around the objects acquired physical significance.

Yet the field did not eliminate every question. What is a field? Is it a physical entity, a mathematical assignment to every point in space, a state of an underlying medium, or a convenient representation of interaction? Field theory gave physics a more powerful language, but the language itself still required interpretation. Einstein's general theory of relativity moved the gravitational problem again. Gravity was no longer treated simply as a force acting within spacetime. It became associated with the geometry of spacetime itself. Matter and energy influenced geometry, and bodies followed paths shaped by that geometry.

Once again, the old concept was not merely discarded. It was reinterpreted within a deeper framework. For ordinary engineering and everyday motion, gravitational force remains extremely useful. For stronger fields, high precision, or cosmological scales, geometry provides a broader description. This historical progression reveals something fundamental about physics. A successful quantity may later become secondary. What one theory treats as a cause, another may treat as a manifestation.

Force appeared fundamental in classical mechanics. In field theories, force became the local effect of a field. In general relativity, gravitational force could be replaced by geometric motion. In quantum field theory, interactions are described through field dynamics and exchange processes rather than classical pushes and pulls. The observable consequences remain. The ontology changes.

This should make us cautious whenever we ask for the "origin of force." Which force are we asking about—the muscular sensation, the Newtonian quantity, the field effect, the geometric interpretation, or the quantum interaction? The word remains the same while its meaning shifts. Newton's dilemma therefore offers a wider lesson. Science can know *how* with great confidence while remaining uncertain about *what*. It can predict a consequence before explaining the ultimate mechanism. It can formulate a law before identifying the deeper structure. This is not a failure. It is often how science progresses. The danger begins only when later generations forget the original boundary and treat a successful description as a completed explanation. Newton's restraint is worth preserving. He gave physics a law strong enough to transform civilization, yet the unanswered question beneath gravity remained open. That openness allowed physics to move forward.

5. Naming Is Not the Same as Creating

When humans name something, two very different events may occur. Sometimes we name an object that exists independently of the name. A mountain does not begin to exist when it receives a label. A newly identified planet does not appear because astronomers assign it a designation. At other times, naming creates a conceptual category. Borders, currencies, legal identities, academic disciplines, and social roles depend partly on shared human agreement. The consequences may be real, but the categories are not natural objects in the same sense as mountains or planets.

Physical quantities occupy an interesting position between these cases. The phenomena they describe are not invented. Objects move, collide, deform, heat, cool, accelerate, and interact whether or not humans observe them. But the quantities through which science organizes these phenomena are constructed. Nature does not divide itself into textbook chapters called force, momentum, pressure, potential, and work. Human intelligence identifies recurring relations, isolates measurable aspects, assigns symbols, defines units, and creates operational rules.

The metre is constructed. Distance is not. The second is defined. Change does not wait for clocks. The newton is a unit created by humans. The measurable relation among mass, acceleration, and interaction is not created by the unit. This distinction allows us to avoid two extremes. The first extreme is naïve realism: assuming every scientific term corresponds directly to a separate entity in nature. The second is careless relativism: assuming that because scientific concepts are human constructions, reality itself is merely invented. Neither position is satisfactory.

Scientific quantities are constructed responses to objective regularities. They are not arbitrary. A scientific concept survives because it connects consistently with observation, measurement, prediction, and experiment. We cannot define force in any way we please and still build functioning machines. Reality constrains the vocabulary. But the vocabulary remains ours. This is similar to a map. A cartographer selects scale, projection, symbols, boundaries, and colours. These are human decisions. Yet the map cannot place mountains and rivers anywhere without losing correspondence with the landscape. The map is constructed. The terrain is discovered.

Scientific quantities work in the same dual way. The structure of nature is discovered. The descriptive system is constructed. Force therefore should not be treated either as a pure invention or as a simple object waiting to be uncovered. It is a scientifically disciplined concept developed to express a stable pattern in the behaviour of matter. This is why the question "Did we discover force or invent it?" cannot be answered with only one word. We discovered that physical systems change in regular and measurable ways. We invented the concept of force to represent one class of those changes. We then refined the concept as our theories developed.

At times, we also reified it—we began speaking as though the label itself were the underlying cause. Reification occurs when an abstraction is treated as a concrete thing. It is easy to do because grammar encourages it. The word *force* is a noun. Nouns sound like entities. We say a force "acts," "causes," "produces," or "exists." Language turns a relationship into an apparent object. The same happens with energy. We say energy flows, is stored, is used, is lost, or is consumed. These expressions are practical, but they can create the impression that energy is a material fluid.

Language enables science, but it also shapes imagination. Once a concept receives a name, symbol, and unit, it acquires authority. It begins to appear as though nature itself contains the category exactly as science defines it. That is why naming must be accompanied by philosophical caution. The label identifies a measurable pattern. It does not necessarily reveal the final structure beneath that pattern. Force may be an excellent description of interaction without being the deepest thing that exists.

6. Why Language Matters in Physics

Physics is often presented as though mathematics protects it entirely from the ambiguity of language. Mathematics certainly reduces ambiguity. An equation can express a relationship more precisely than ordinary speech. But equations do not interpret themselves. Scientists still choose the words used to explain what the symbols mean, and those words influence the way generations imagine reality.

Consider the phrase "force acts on an object." The grammar contains a subject, a verb, and an object. Force becomes the acting subject, and the material body becomes the receiver. The sentence quietly suggests that force is a kind of agent. Now consider a different statement: "The object's motion changes according to the local physical conditions and its relation to surrounding systems." This sounds less direct, but it does not turn force into an independent actor. It presents motion change as a consequence of interaction. Both statements may lead to the same calculation, but they produce different ontological pictures.

Language matters because scientific education does more than teach formulas. It builds an internal model of reality. A student repeatedly told that forces push and pull objects may imagine invisible hands acting throughout nature. A student told that mass bends spacetime may imagine a flexible sheet, even though the popular image is only an analogy. A student told that particles exchange messenger particles may imagine tiny objects being thrown between them. Analogies are useful, but they can harden into assumed reality.

The history of physics contains many terms whose original meanings shifted dramatically. The word *atom* once suggested something indivisible, yet physics later discovered internal structure. The word *particle* suggests a tiny solid object, yet quantum particles do not behave like miniature stones. The word *wave* suggests a disturbance in a medium, but the wavefunction in quantum mechanics is not simply an ordinary physical wave. The word *field* originally helped describe distributed influence, but modern fields carry mathematical and ontological meanings far beyond the everyday idea.

Language remains necessary, but it can never be perfectly neutral. Even the phrase "laws of nature" can mislead. Human laws are commands imposed by authorities. Physical laws do not order nature to behave. They summarize regularities observed in behaviour. The universe does not obey Newton. Newton described a pattern found within the universe. This distinction may sound obvious, yet everyday scientific speech often reverses it. We say objects "obey laws," energy "wants" to minimize, nature "chooses" the easiest path, or particles "know" whether they are being observed. Such expressions are shortcuts. Scientists usually understand their limited role. But when these phrases leave specialist contexts and enter public culture, the metaphor may be mistaken for mechanism.

The danger increases when speculative physics is presented using familiar nouns. Once something is called a particle, field, dimension, or universe, it begins to sound physically established even when it remains hypothetical. A name gives cognitive solidity. This is why scientific language must retain markers of certainty. Observed is not the same as inferred. Inferred is not the same as modelled. Modelled is not the same as confirmed. Mathematically possible is not the same as physically real.

Force itself offers a perfect case study. The word can refer to muscular effort, a measured mechanical quantity, a field effect, an interaction term, or an apparent effect arising in an accelerating frame. When one word carries several layers, confusion becomes almost unavoidable unless the context is made explicit. This book therefore treats physical quantities as a manifestation vocabulary. The phrase does not deny their reality or usefulness. It places them correctly within the hierarchy of knowledge. Reality behaves. Manifestations become observable. Quantities describe selected aspects of those manifestations. Equations express relationships among quantities. Theories organize those equations. Interpretations explain what the theories may mean. Language sits between every layer. It enables understanding, but it can also quietly convert description into ontology. To think clearly about physics, we must learn not only how to use its vocabulary, but also when to look beyond it.

7. From Force to Energy Gradients and Fields

If force is not necessarily a fundamental object, what is the deeper alternative? Modern physics has already moved much of the way toward an answer. In many situations, a force can be related to a change in potential energy across position. Objects respond because the surrounding physical condition is not uniform. There is a difference, slope, gradient, curvature, or field structure.

A ball rolls downhill not because "downhill force" exists as a separate substance, but because the system occupies a configuration in which motion toward a lower potential becomes possible. A charged particle changes motion in an electromagnetic field. A pressure difference drives fluid flow. A temperature difference drives heat transfer. A concentration difference drives diffusion. Across physics, change frequently emerges from non-uniformity. Something differs across space, time, configuration, or state. The system responds. Force becomes one way of expressing that response.

This does not mean that every physical process can be reduced carelessly to a simple energy gradient. Physics contains multiple frameworks, and the language of fields, symmetries, geometry, quantum states, and interactions cannot be compressed into one slogan. Still, the general movement is revealing. Science increasingly seeks the structure that makes the observed response possible. Instead of stopping at "a force acts," it asks: what field exists, what potential changes, what symmetry is involved, what geometry governs the path, what interaction changes the state, what deeper conditions generate the measurable consequence?

This is already broader than classical force-centred physics. It suggests that matter does not need to experience force as a conscious entity. Matter responds according to its structure and its relation to the surrounding physical environment. The language of response may be more fundamental than the language of action. A magnet does not consciously pull iron. The magnetic configuration produces conditions under which the iron responds. Earth does not extend an invisible hand toward the Moon. The Earth–Moon system evolves according to the gravitational structure in which both are embedded. A spring does not decide to restore its original length. Its material configuration changes when displaced, and the resulting internal energy distribution produces a measurable response.

In this view, force belongs to the layer of manifestation and calculation. The deeper physical layer consists of structured relations and energy dynamics. This is where a substrate-based question naturally appears. If fields, potentials, matter, and interactions are physically real in some sense, within what broader foundation do they exist? Are fields final? Is spacetime final? Is vacuum truly empty? Or are these themselves manifestations of a deeper universal structure?

Modern physics has not produced one universally accepted final answer. Quantum field theory gives the vacuum physical properties. Relativity gives spacetime dynamical structure. Cosmology attributes large-scale behaviour to forms of energy that remain incompletely understood. The old idea of empty nothingness has already become difficult to maintain. Universal Energy Dynamics begins from the proposal that reality is grounded in a structured universal substrate, referred to as Structured Vacuum Energy, or SVE. Matter is then approached not as something placed inside an empty background, but as a stable or organized form within the substrate. Fields and interactions become expressions of underlying energy dynamics. Physical quantities describe the manifestations of those dynamics.

At this stage, this should be treated as a conceptual doorway, not as a conclusion the reader must accept. The important shift is simpler: force may not be the starting point of physical reality. It may be the measurable language of a deeper response. That possibility does not weaken mechanics. Engineers can continue calculating forces. Students can continue drawing free-body diagrams. Satellites can continue being placed in orbit using established equations. The practical description remains valid. But beneath the successful calculation, another question becomes visible: what is reality doing before we call the result force?

Closing Reflection: The Name and the Event

Return once more to the wall. Place your hand against it and push. Your body produces effort, your muscles consume energy, the skin registers pressure, the wall resists visible displacement, the floor supports your body, and Earth participates in the interaction. Internal structures across your arm, the wall, and the ground respond continuously. The word *force* gathers this complexity into a usable concept. That is its power. It allows a teacher to explain, a student to calculate, an engineer to design, and a scientist to compare. Without such concepts, physics would remain trapped in personal sensation.

But the word is not the event. The symbol is not the interaction. The equation is not the complete architecture of what occurs. We did not invent the fact that motion changes, matter resists deformation, objects collide, gradients produce responses, and systems exchange energy. We discovered those regularities. But we did invent the language through which we isolate, measure, and relate them. Force belongs to that language. It is neither a mere fiction nor necessarily an ultimate substance. It is a disciplined scientific description of recurring consequences within physical interaction.

This is why asking whether force was discovered or invented leads to a more mature answer: the phenomenon was discovered, the quantity was formulated, the name was invented, the relationship was measured, and the underlying reality remains open to deeper explanation. Science becomes stronger when it remembers these distinctions. It can use force without worshipping the concept. It can trust the equation without pretending that the equation has exhausted reality. It can preserve the precision of classical mechanics while still asking what fields, potentials, geometry, matter, and energy ultimately represent.

Perhaps the most important scientific question is not always *what force is acting*. Sometimes it is *what is happening here that our language has chosen to call force*.

Chapter 2 — Nature Does Not Know Physics

Before the Equation Arrives

A stone falls from a tree. The event appears simple. The stone separates from the branch, moves downward, accelerates, strikes the ground, and comes to rest. A physicist may describe this event using mass, height, acceleration, gravitational potential energy, kinetic energy, momentum, impact force, and deformation. A more advanced analysis may include air resistance, rotation, heat, sound, molecular structure, and the curvature of spacetime. Each description adds precision. Yet the stone knows none of them.

It does not know its mass in kilograms. It does not know that its acceleration is approximately related to the gravitational conditions near Earth's surface. It does not calculate its trajectory. It does not convert potential energy into kinetic energy. It does not consult Newton before moving or Einstein before falling. It simply participates in reality. The equations arrive later. This does not make the equations false. On the contrary, their ability to describe and predict the event is one of the greatest achievements of human intelligence. But the order must remain clear.

First, something happens. Then it is observed. Then selected aspects are measured. Then quantities are introduced. Then relationships are expressed mathematically. Then a theory is built around those relationships. Nature does not begin with the theory. The theory begins with nature.

This distinction seems obvious, yet it is frequently lost. Scientific education often begins with the finished equation rather than the event from which it emerged. Students may come to believe that objects behave as they do because a law commands them. We say that planets obey Newton's laws. We say that light obeys the laws of optics. We say that energy obeys conservation laws. But nature does not obey in the human sense. No authority issues instructions to matter. Physical laws are not regulations imposed upon the universe. They are disciplined human descriptions of regular patterns found within it. A law does not cause the pattern. The pattern makes the law possible.

This chapter begins from a simple but far-reaching proposition: nature does not know physics. Physics is how conscious intelligence organizes what nature does. Once this is understood, physics becomes both stronger and more open. Stronger, because its equations can be respected for what they genuinely achieve. More open, because no successful description needs to be mistaken for reality's final form. The universe does not live inside our theories. Our theories live inside the universe.

1. Reality Comes Before Description

Reality does not wait for a description in order to exist. Stars formed before astronomy. Objects fell before mechanics. Lightning occurred before electricity was understood. Heat moved before thermodynamics. Living organisms evolved before biology. Atoms interacted before atomic theory. The sequence is always the same. Nature acts first. Human knowledge follows.

This may sound trivial, but it carries an important philosophical consequence. Scientific descriptions are necessarily secondary. They are responses to reality, not the origin of reality. A physical event is not created by the language used to describe it. When water boils, molecular motion, pressure, phase transition, heat transfer, and surrounding conditions are involved. Yet none of those scientific terms produces the boiling. They identify and organize aspects of what is already happening.

The same event can also be described differently depending on the level of analysis. A cook may say that the water is ready. A child may say that bubbles are rising. A thermodynamic description may refer to temperature, pressure, latent heat, and phase equilibrium. A molecular description may refer to distributions of molecular energy. A quantum description may eventually consider interactions at a deeper microscopic level. These descriptions do not necessarily contradict one another. They operate at different scales and for different purposes. Reality is richer than any one vocabulary.

This becomes especially clear when science divides nature into disciplines. The falling leaf may be studied through mechanics, fluid dynamics, thermodynamics, biology, chemistry, or mathematics. The leaf itself does not divide its behaviour according to university departments. The separation belongs to human organization. Nature is continuous. Knowledge is departmental. We construct categories because the total complexity of reality cannot be studied all at once. We isolate motion from heat, heat from chemistry, chemistry from biology, and biology from psychology. This makes investigation manageable.

But isolation can produce an illusion. Once a discipline becomes well established, its concepts may appear to be built into reality itself. We begin speaking as though force, charge, entropy, information, life, and consciousness exist in perfectly separate domains. In practice, the boundaries are often porous. A moving machine generates heat. Heat changes material properties. Material changes influence mechanical behaviour. Mechanical failure affects human decisions. Human decisions alter technological systems. Reality does not respect the borders of our textbooks.

This is why the distinction between reality and description matters. A description selects. To describe anything is to leave something out. A mechanical model may ignore colour, smell, history, and biological meaning. A thermodynamic model may ignore the exact trajectories of individual molecules. A statistical model may ignore the identity of each component. A cosmological model may ignore local details in order to represent large-scale structure. Selection is not a defect. It is the condition of usefulness. A model that included everything would be as complicated as reality itself. It would cease to function as a model.

Scientific description therefore depends on reduction. It identifies certain variables as relevant and treats others as negligible under specified conditions. The resulting theory can be extraordinarily powerful without being complete. This is the first principle we must retain: a description succeeds by capturing what matters for a particular question, not by reproducing the whole of reality. Theories are therefore not photographs of the universe. They are structured ways of relating selected manifestations. Reality comes first. Description follows. The mistake begins when the description is promoted into the thing itself.

2. Does the Universe Know Newton's Laws?

The phrase "laws of nature" is so familiar that its metaphorical character is rarely noticed. Human laws are written, announced, enforced, interpreted, and sometimes violated. They presuppose a lawmaker, a subject, and a system of authority. Physical laws are entirely different. A planet cannot violate the law of gravitation and later be punished. A photon cannot disobey electrodynamics. A gas does not study thermodynamics before expanding. Physical laws do not command nature. They summarize regularity.

This distinction became especially important after Newton. His laws of motion and universal gravitation achieved something remarkable: they expressed an enormous range of physical behaviour through a compact mathematical framework. The same mechanics could describe falling objects, projectiles, pendulums, machines, moons, and planets. Nature appeared mathematically unified. This success encouraged a powerful image of the universe as a machine governed by fixed laws. If the position and motion of every particle were known, perhaps the entire future could be calculated.

The image was scientifically productive, but it also encouraged a subtle reversal. Instead of saying that Newton's equations describe regularities found in motion, people began saying that motion occurs because objects obey Newton's laws. The law became the apparent cause. But what would that mean? Does an object somehow contain the equation $F = ma$? Does the universe perform calculations before events occur? Does matter read mathematical instructions? Clearly not. The equation belongs to the human description of the relation. The object participates in the physical process.

This does not reduce the equation's authority. A bridge designed with poor mechanics may collapse. A spacecraft launched with incorrect calculations may miss its target. Nature constrains the mathematics severely. But constraint is not obedience. Newton's laws succeed because they correspond closely to stable physical patterns within a certain domain. They are especially effective when speeds are not close to the speed of light, gravitational fields are not extreme, and quantum effects are negligible. Outside those conditions, broader theories become necessary.

Relativity does not mean that Newton was simply wrong. It means Newtonian mechanics is an approximation within a larger structure. This is an important clue. If the universe literally obeyed Newton's laws as final commands, why would those laws require modification at high speeds, strong gravity, or microscopic scales? The answer is that laws are descriptions with domains of validity. A physical law is not necessarily a final statement of what reality is. It may be an effective relationship that becomes increasingly accurate under specified conditions.

Consider the motion of Mercury around the Sun. Newtonian gravitation explained planetary motion with extraordinary success, yet a small discrepancy remained in Mercury's orbit. The discrepancy was not large enough to destroy Newtonian mechanics, but it signalled that the description was not complete. Einstein's general relativity later accounted for the behaviour through a different framework. Gravity was no longer treated merely as a force between masses. It became associated with spacetime geometry. The planet did not change its behaviour when Einstein published his theory. Human description changed. Mercury had always moved as it moved. Newton captured most of the pattern. Einstein described it more deeply.

This episode reveals what scientific progress often means. Nature does not move from one law to another. Scientists move from one model to another. The universe does not learn physics. Physics learns the universe.

3. Maps Are Not Territories

A map can be accurate without being the landscape. It may show roads, rivers, borders, elevations, railways, or population density. Each map selects different features depending on its purpose. A road map need not include every tree. A geological map need not show every shop. A weather map need not represent property ownership. No one criticizes a map for omitting irrelevant details. Its value lies in useful correspondence.

Scientific theories function in a similar way. A theory identifies variables, relationships, limits, and conditions. It allows us to navigate a class of phenomena. It tells us what to expect, what to measure, and how one change relates to another. But like maps, theories are selective. A mechanical model of a pendulum may treat the bob as a point mass, the string as massless, the pivot as frictionless, and the air as absent. None of these conditions is perfectly true in the physical world. Yet the model works.

Why? Because the ignored details may be small enough that the simplified structure captures the dominant behaviour. This is not deception. It is disciplined idealization. Physics depends on such idealizations. Perfectly rigid bodies do not exist. Point particles are conceptual simplifications. Frictionless surfaces are approximations. Isolated systems are rarely completely isolated. Uniform fields are limited constructions. Nevertheless, theories built from these concepts can predict real events with high accuracy.

This tells us something important: a model does not need to resemble reality in every detail to be effective. It needs to preserve the relevant structure. A subway map is a good example. It may distort geographical distances and angles, yet it helps passengers reach the correct station. Its practical truth lies in preserving connectivity rather than physical scale. Likewise, a physical theory may preserve the relations necessary for prediction without providing a literal picture of what reality looks like underneath.

This becomes crucial in modern physics, where mathematical structures often exceed ordinary visualization. Quantum mechanics predicts experimental outcomes with extraordinary precision, but physicists continue to disagree about what the mathematical formalism says reality is doing between measurements. The map works. The nature of the territory remains debated.

This is not unusual. Even in classical physics, the success of a model does not guarantee that its conceptual imagery is final. The old Ptolemaic system of astronomy used circles and epicycles to reproduce planetary positions. It was not merely irrational superstition. It was a serious predictive framework for its time. Its weakness was not that it produced no useful results, but that its underlying structure became increasingly complicated and eventually gave way to a more coherent astronomical picture.

A model may therefore succeed for several reasons. It may genuinely capture a deep structure. It may approximate a deeper theory under limited conditions. It may organize observations effectively without representing the underlying mechanism literally. It may even fit existing data while later proving conceptually misleading. This is why predictive success should not be confused with ontological completion. A map can guide you accurately while remaining silent about the soil beneath the road. A theory can calculate an orbit without explaining what gravity ultimately is. An equation can predict a spectrum without revealing what a quantum state fundamentally represents.

Scientific maturity requires us to ask two different questions: does the model work, and what, if anything, does the model tell us exists? The first may be answered experimentally. The second often requires interpretation. When these questions are collapsed, the map begins to replace the territory.

4. Why Simplified Theories Work

If scientific theories are incomplete, selective, and sometimes idealized, why do they work so well? This is one of the most fascinating questions in physics. The answer begins with regularity. Reality is not random in every respect. Similar conditions often produce similar outcomes. Objects near Earth fall in predictable ways. Materials respond systematically to temperature and stress. Electrical circuits exhibit stable relations. Light refracts according to measurable patterns. Planetary motion remains sufficiently regular for eclipses to be predicted. Without this consistency, science would be impossible.

Theories succeed because they capture recurring structure. But the relationship between a theory and reality is rarely one-to-one. The theory often focuses on a limited scale where many details can be ignored. For example, the motion of a large object can often be described without tracking every atom within it. Engineers calculate the load on a beam without solving the quantum state of every electron. Meteorologists model large-scale atmospheric behaviour without following every air molecule individually. At different scales, different variables become useful. This is the principle of effective description.

A theory can be valid at one level without being fundamental at all levels. Thermodynamics describes temperature, pressure, volume, and entropy without needing to specify every microscopic motion. Its laws remain powerful even though temperature and pressure emerge from underlying collective behaviour. Fluid mechanics treats water as a continuous medium, although water consists of molecules. The continuum model succeeds because molecular details can often be averaged at the scale of interest. Classical mechanics treats objects as bodies with mass and motion, even though matter has quantum structure. For ordinary speeds and dimensions, the quantum details may not significantly alter the calculation.

Theories therefore succeed partly because reality contains layers of stable organization. A higher-level pattern can remain reliable even when it arises from deeper processes. This should change how we think about scientific truth. A theory does not need to be the final theory to be true within its range. Newtonian mechanics is true enough to design vehicles, buildings, machines, and many spacecraft trajectories. Relativity does not erase these achievements. It clarifies the conditions under which Newton's framework is an approximation. Similarly, classical thermodynamics remains valid even though its quantities arise from molecular dynamics. The emergence of a deeper explanation does not destroy the higher-level description.

This layered structure helps explain why science can progress without constantly discarding everything that came before. New theories often contain older theories as limiting cases. The older map remains useful for a familiar region. The newer map covers more territory. Yet there is another reason theories work: human beings design them to match observation. A theory is not accepted simply because it is imaginative. It must survive comparison with experiment. Failed predictions force revision. Contradictions expose hidden assumptions. New instruments reveal deviations that were previously invisible. Scientific theories are shaped by reality through repeated correction.

That is why they can be both human constructions and objectively constrained. We invent the framework. Nature decides whether it survives. This relationship is neither pure discovery nor pure invention. The regularity is discovered. The model is constructed. Its adequacy is tested. Its domain is established. Its meaning is debated. Theories work because they are disciplined by consequences. But a theory's success within its domain does not prove that its language corresponds directly to the deepest structure of reality. It proves that the selected relationships are reliable enough to describe and predict the phenomena under consideration. That is a major achievement. It is not the same as final ontology.

5. When Success Creates Certainty

A successful theory gradually becomes invisible. At first, it may be controversial. Scientists debate its assumptions, test its predictions, question its meaning, and compare it with competing explanations. If it survives, it becomes standard. It enters textbooks, classrooms, engineering practice, and institutional training. Eventually, students encounter the theory without encountering the original uncertainty. The struggle disappears. Only the finished framework remains. This produces efficiency, but it can also create false certainty.

Consider the transition from classical mechanics to relativity. For centuries, Newtonian concepts of absolute space and time appeared natural. They were not treated merely as convenient assumptions; they became part of the scientific imagination. Then problems involving electromagnetism, motion, and the speed of light exposed tensions. Einstein's work did not simply add another formula. It altered the meaning of space, time, simultaneity, mass, and gravity. Concepts once treated as obvious turned out to be provisional.

The same pattern appears in quantum physics. Classical intuition assumed that physical objects possess definite properties and follow definite trajectories. Quantum experiments forced physics to adopt a mathematical structure that did not fit this picture cleanly. The theory worked extremely well. Its interpretation remained contested. Einstein, Bohr, Schrödinger, Heisenberg, and others did not merely argue over calculations. They disagreed about what the successful calculations implied about reality.

This is a crucial distinction. A theory can achieve operational agreement while leaving ontological disagreement unresolved. Physicists may agree on how to calculate an experimental result while disagreeing about what exists, what happens during measurement, whether probabilities are fundamental, or whether the mathematical state represents reality or knowledge. The public often does not see this distinction. Once a theory is associated with successful experiments and advanced mathematics, every interpretation surrounding it may appear equally established.

This is where physics can drift toward metaphysics without clearly announcing the transition. The drift does not occur because physicists are careless. It occurs because every theory invites a picture of reality, and human intelligence naturally seeks such pictures. But a picture may be more speculative than the calculation. For example, a mathematical framework may permit multiple interpretations. One interpretation may speak of collapse, another of branching worlds, another of hidden variables, and another of relational states. The observed results may remain the same. The ontological stories differ. Which story is "the physics"? The answer is not always simple.

Institutions also play a role. Once a framework becomes dominant, research programmes, funding structures, curricula, journals, and careers may develop around it. This does not make the framework false. But it can make alternative questions harder to pursue. Scientific communities need standards. Without them, knowledge fragments into unsupported claims. Yet standards can also harden into assumptions if the distinction between evidence and interpretation is neglected.

Success therefore creates a paradox. The more useful a theory becomes, the more likely people are to mistake usefulness for completeness. The more accurate its predictions, the more its conceptual language appears to describe reality directly. The more deeply it enters technology, the less often its foundations are questioned. This is why foundational inquiry remains necessary. Not to undermine established science, but to prevent established descriptions from becoming unquestioned mythology.

A mature science should be able to say: this theory works, this is where it works, this is what it predicts, this is what has been observed, this is what remains interpreted, and this is what remains unknown. Such clarity protects both science and imagination.

6. The Difference Between Prediction and Explanation

Prediction is one of the strongest tests of a scientific theory. If a model consistently forecasts outcomes that experiments later confirm, confidence in the model increases. Prediction shows that the theory captures something real about the structure of events. But prediction and explanation are not identical. A calendar can predict an eclipse without explaining gravity. A mathematical curve can fit data without revealing the mechanism generating it. An algorithm can forecast behaviour without understanding the system in any conscious sense. Similarly, a physical equation can accurately predict an outcome while leaving open the question of what fundamentally produces it.

Newtonian gravitation predicted planetary motion with extraordinary success, yet the mechanism of gravitational action remained unclear. Quantum mechanics predicts probabilities with astonishing precision, yet its interpretation remains philosophically difficult. Statistical mechanics explains thermodynamic behaviour through large numbers of particles, but the transition from microscopic reversibility to macroscopic irreversibility still raises deep questions about time and entropy. Cosmological models can fit observations while containing components whose fundamental nature is not yet understood. Science frequently knows the relation before it knows the underlying reality. This is not embarrassing. It is the normal condition of developing knowledge.

The danger lies in treating predictive success as though it automatically closes the explanatory question. Suppose a theory predicts that a particle will appear in a detector with a particular probability. The prediction is confirmed. What has been established? At minimum, the mathematical relation corresponds reliably to observed outcomes. But has the experiment proved one specific interpretation of the mathematical state? Not necessarily. The data may support the formalism while leaving several ontological readings possible.

This distinction becomes even more important when scientific language reaches the public. A phrase such as "particles exist in multiple places at once" may sound like a direct observation. In reality, it may be an interpretation of a formal structure and experimental pattern. The experiment is real. The mathematical prediction is real. The interpretation requires care. Good science should distinguish all three.

Explanation itself also has layers. One theory may explain a phenomenon by relating it to a deeper law. Another may explain it by showing how it emerges from microscopic behaviour. Another may explain it through geometry, symmetry, or conservation. But every explanation eventually reaches a point where it rests on something not further explained within that framework. Why does an object accelerate? Because of a force. Why does the force arise? Because of a field. Why does the field have that structure? Because of equations, sources, symmetries, or boundary conditions. Why do those structures exist? At some point, the explanation ends—or passes into a deeper theory.

Scientific progress pushes that boundary outward. It does not necessarily eliminate the boundary. This is why explanation should not be confused with renaming. Calling an effect "gravitational," "quantum," or "field-mediated" may classify it successfully, but classification alone is not a complete explanation. A word can mark the location of a mystery without resolving it. The proper role of theory is therefore not to pretend that every successful description has become final reality. Its role is to organize evidence, predict consequences, expose relationships, and open new questions. A strong theory does not end inquiry. It makes deeper inquiry possible.

7. What Nature Is Doing Beneath Our Theories

If nature does not know physics, what does it do? The answer must be stated carefully. Nature does not "do" in the intentional human sense. It does not plan, calculate, obey, or interpret. These are metaphors borrowed from conscious activity. Yet reality continuously changes, organizes, interacts, stabilizes, transforms, and produces consequences. Matter forms. Fields vary. Energy redistributes. Systems oscillate. Structures emerge. Patterns persist.

Some configurations remain stable for long periods. Others dissolve rapidly. Some interactions produce motion. Others produce heat, light, sound, deformation, or new material forms. Physics observes these manifestations and constructs a vocabulary around them. Force describes one relation. Energy describes another. Momentum, pressure, temperature, charge, entropy, and field each identify measurable aspects of the broader dynamics. But the quantities should not be mistaken for the process itself.

This suggests a hierarchy: Reality → Dynamics → Manifestations → Observation → Measurement → Physical quantities → Equations → Theories → Interpretations.

The lower layers depend on the higher ones, but not the reverse. Reality does not depend on the equation. The equation depends on the observed regularity. This hierarchy is central to the broader view developed in this book. It allows us to preserve everything useful in physics while questioning what has been mistaken for fundamental. A force equation may remain valid even if force is treated as a manifestation variable. A thermodynamic law may remain valid even if temperature is emergent. A field equation may remain valid even if fields arise from a deeper substrate.

Scientific progress often works precisely this way. A quantity remains useful while its ontological status changes. In Universal Energy Dynamics, the deeper proposal is that matter, fields, and physical manifestations arise through the structured dynamics of a universal substrate referred to as Structured Vacuum Energy, or SVE. This idea should not be accepted merely because it provides a broad picture. It must be judged by the same discipline applied to every other framework. What does it clarify? What does it predict? How does it differ from existing descriptions? Can its claims be tested? Where does interpretation end and measurable consequence begin?

The purpose of introducing SVE here is not to replace one unquestioned vocabulary with another. That would repeat the same mistake. The purpose is to reopen the foundational question: what kind of reality must exist for fields, matter, energy, motion, and measurable quantities to appear at all? Current physics offers powerful models of these phenomena. Yet the very existence of multiple successful frameworks suggests that no single vocabulary should be treated casually as the final substance of nature.

Classical mechanics speaks through forces. Relativity speaks through geometry. Quantum theory speaks through states, amplitudes, operators, and probabilities. Thermodynamics speaks through macroscopic variables. Each captures a real domain. Each selects a level. Each has limits. Nature contains all the phenomena these theories describe, but it does not divide itself according to our preferred formulation. Reality is one. Descriptions are many.

The deeper task of physics may therefore be not only to improve calculation, but to understand how these descriptive layers arise from a common underlying structure. That is where physics moves beyond quantities without abandoning them. It begins to study manifestation itself.

Closing Reflection: The Universe Without a Textbook

Imagine the universe before the first human observer appeared. Stars formed and collapsed. Planets moved. Radiation travelled. Matter organized into increasingly complex structures. There were no equations written anywhere. No units. No textbooks. No symbols. No scientific institutions. No distinction between mechanics, thermodynamics, relativity, or quantum theory. Reality continued without our vocabulary.

Then conscious beings emerged within that reality. They observed repetition. They compared events. They measured differences. They developed numbers, geometry, instruments, quantities, laws, and theories. Through these tools, the universe began to be described from within itself. This is one of the most extraordinary events in existence: matter organized into living intelligence capable of modelling the reality from which it emerged. But the model remains inside the reality. It cannot become larger than the territory it describes.

Nature does not know Newton's laws. It does not know Einstein's equations. It does not know the names of particles, forces, fields, or dimensions. These belong to human knowledge. What nature contains are the stable patterns that make such knowledge possible. Theories succeed because reality is ordered enough to be described. They succeed because human intelligence can isolate relevant structures, express them mathematically, and test them against consequences. But success does not turn a theory into reality itself.

Newtonian mechanics remains powerful without being final. Relativity deepens gravity without exhausting every question. Quantum theory predicts with remarkable accuracy while its interpretation remains open. Every theory is a map drawn from within the territory. Some maps are rough. Some are extraordinarily precise. Some overlap. Some apply only to particular regions. Some reveal structures that earlier maps could not show. But none should be confused with the landscape itself.

The proper scientific attitude is therefore neither blind certainty nor careless doubt. It is disciplined confidence. Use the theory where it works. Respect the evidence supporting it. Recognize the assumptions within it. Mark the boundary between observation and interpretation. Remain open to a deeper map. Nature does not know physics. But through physics, we are learning to know nature. And perhaps the greatest progress will come when we stop asking whether reality obeys our theories and begin asking why our theories capture only certain layers of what reality is already doing.

Chapter 3 — The Hidden Meaning of Physical Quantities

When a Number Replaces an Experience

A child places a hand on a steel railing under the afternoon sun and immediately pulls it back. “It is hot,” the child says. An adult may agree. A scientist may place a thermometer against the railing and report a temperature. An engineer may ask how quickly the metal absorbs and transfers heat. A physicist may describe the average energy distribution among particles within the material. Each response refers to the same event, but they do not describe the same layer.

The child reports an experience. The thermometer reports a measurement. The physicist introduces a quantity. The theory explains a relationship. The metal itself does none of these things. It does not know that it is hot. It does not know that its temperature is fifty degrees. It does not know the unit in which the reading is expressed. It simply exists in a particular energetic condition and interacts with the hand accordingly. The quantity comes afterward.

This pattern is repeated throughout physics. A body resists acceleration, and we introduce mass. A material system exchanges energy, and we introduce temperature. Molecules collide with a surface, and we introduce pressure. Matter interacts electromagnetically, and we introduce charge. A system evolves toward configurations of greater statistical accessibility, and we introduce entropy. The quantities are powerful because they condense complex physical behaviour into measurable variables. They enable prediction, comparison, engineering, and communication. Without them, modern science would collapse into vague description.

But their success creates a hidden temptation. Once a quantity receives a name, symbol, unit, and equation, it begins to appear like an independent ingredient of reality. We may start imagining that objects contain mass in the way a bottle contains water. We may speak as though temperature is a substance stored inside matter, pressure is a thing pushing outward, charge is a coloured mark attached to a particle, or entropy is an invisible agent driving disorder. The language is useful. The picture may be misleading.

Physical quantities do not necessarily tell us what reality is made of. They tell us how selected aspects of reality can be measured and related. That distinction gives physical quantities their hidden meaning. They are not merely numbers. They are not merely properties. They are conceptual bridges between manifestation and measurement. A quantity begins when conscious intelligence notices a stable difference, develops a method to compare it, defines a scale, and embeds the result within a theoretical structure. The physical event exists independently. The quantity belongs to the scientific response.

This chapter examines that relationship more closely. What does mass actually mean? Where does temperature exist? Is pressure inside a gas, or is it a description of collective interaction? What does charge name? Why is entropy so difficult to visualize? And what changes when we stop treating these quantities as pieces of reality and begin treating them as the manifestation vocabulary of physics?

1. A Physical Quantity Is Not Just a Number

A physical quantity is often introduced as something that can be measured and expressed by a numerical value together with a unit. That definition is correct, but incomplete. A quantity is not created merely by attaching a number to an event. It requires a conceptual decision about what is being compared, how the comparison will be made, which differences matter, and what standard will be used.

Before measuring length, we must decide what counts as separation between two points. Before measuring time, we must identify repeatable change. Before measuring temperature, we must establish a scale that connects observable effects to thermal condition. Before measuring force, we must define how interaction relates to motion or deformation. Measurement therefore does not simply collect ready-made numbers from nature. It organizes reality according to a chosen structure.

Suppose three stones lie on the ground. One can be compared by size, mass, colour, temperature, density, shape, chemical composition, age, or market value. The stones do not announce which variable is most fundamental. The choice depends on the question being asked. If we want to build a wall, mass and shape may matter. If we want to identify the mineral, composition matters. If we want to understand heat transfer, temperature and thermal conductivity matter. A physical quantity is therefore a disciplined selection. It isolates one relation from the larger event.

This is why the same object can possess many quantities without being reducible to any one of them. A piece of metal may have mass, volume, density, temperature, charge, conductivity, momentum, internal energy, and entropy. None of these quantities is the metal itself. Together they form a scientific profile. The profile can be extraordinarily precise. It may allow us to predict how the metal will behave under heat, stress, current, pressure, or motion. But it remains a representation assembled from selected variables.

This distinction is easy to overlook because quantities feel objective. The number appears to be simply there. A weighing scale displays a value. A thermometer gives a reading. A voltmeter measures potential difference. The instrument seems to extract information directly from the object. Yet every instrument embodies a theory. A thermometer works because some physical property changes consistently with thermal condition. A weighing scale converts deformation, balance, or electrical response into a numerical reading. A pressure gauge responds mechanically or electronically to interactions within a fluid.

The instrument does not reveal reality without interpretation. It produces a response that has been calibrated according to a conceptual framework. Measurement is therefore never completely theory-free. It depends on definitions, standards, procedures, and assumptions. This does not make measurement subjective or arbitrary. Reliable measurement is possible precisely because nature responds consistently. Different observers using properly calibrated instruments can obtain comparable results. Objectivity is achieved through disciplined agreement with recurring physical behaviour.

But the quantity remains a bridge. On one side lies the physical system. On the other lies the observer, instrument, scale, unit, and theory. The number is produced at the interface. This gives us a useful principle: a physical quantity is not a piece of nature extracted intact. It is a scientifically constructed relation between an observable manifestation and a defined method of comparison. The manifestation constrains the quantity. The quantity organizes the manifestation. Physics becomes precise through this relationship. It becomes confused only when the relation is mistaken for an independent substance.

2. Mass: Substance, Resistance, or Relation?

Mass appears to be one of the most obvious quantities in physics. An object is heavy or light. A stone is harder to move than a ball. A loaded vehicle responds differently from an empty one. From early experience, mass seems to express how much "stuff" an object contains. This intuition is useful, but physics complicates it quickly.

Mass enters science in more than one way. In mechanics, inertial mass describes resistance to acceleration. A larger mass changes its motion less under the same applied interaction. In gravitation, mass determines participation in gravitational relationships. In relativity, mass is connected to energy and the structure of spacetime. In particle physics, the origin of particle mass is discussed through interactions with fields. The familiar word remains, but its theoretical role changes.

This should make us cautious about treating mass as a simple substance stored inside matter. When a scale measures mass, it does not directly inspect the inner "amount of matter." It registers a response. On Earth, most household scales measure force associated with gravitational interaction and then convert that force into a mass reading under assumed local conditions. The instrument infers the quantity from a physical consequence. Similarly, inertial mass is known through the relationship between interaction and acceleration. We do not see mass directly. We observe how the object responds.

Mass therefore describes a stable aspect of material behaviour. It tells us something important and objective, but what it tells us is relational. The history of physics makes this even clearer. Newton treated mass as a fundamental measure of matter. Classical mechanics used it with enormous success. Einstein later showed that measurements of energy, momentum, time, and motion required a broader framework at high speeds. The relationship between mass and energy altered the older picture of matter as an independent, permanent substance.

Later particle physics deepened the issue again. The mass of composite objects does not arise only from adding the masses of their constituent particles. Much of the mass of ordinary matter is associated with internal energy and interactions. The object appears solid and stable, yet the quantity we call mass reflects organized dynamics within it. This does not mean mass is unreal. It means its reality may not be substance-like. Mass may be better understood as a manifestation of how an organized physical system resists change, participates in gravitational structure, and stores or expresses energy.

The quantity captures a persistent relation across many situations. But the word encourages reification. Because mass is a noun, we imagine it as something the object possesses independently, like a hidden material filling. The scientific equations do not require this picture. They require that measurable relations remain consistent. A spacecraft can be designed using mass without resolving the ultimate ontology of mass. A planet's motion can be calculated without claiming that mass is a visible ingredient.

This distinction matters for a substrate-based view of reality. If matter is not a collection of self-existing solid substances but an organized form of deeper energy dynamics, then mass may describe the stability and responsiveness of that organization. In Universal Energy Dynamics, one could approach mass as a manifestation of structured energy held within a persistent material configuration of the universal substrate. That is not yet a complete physical theory of mass. It is a directional interpretation. Its value lies in changing the question.

Instead of asking, "How much mass does the object contain?" we may ask, "What organized physical structure produces the stable response that we measure as mass?" The quantity remains. The ontology becomes deeper. Mass is therefore not merely "how much matter." It is a measurement vocabulary for a set of stable physical responses whose underlying structure continues to be investigated.

3. Temperature: A Quantity That No Particle Possesses

Temperature provides one of the clearest examples of a physical quantity that is real without existing as a separate microscopic object. A cup of water can be assigned a temperature. But does each individual molecule possess the temperature of the cup? Not in the same way. The molecules have different motions, energies, positions, and interactions. Temperature emerges from the collective statistical condition of the system.

This means temperature belongs to the organized behaviour of many constituents, not to an isolated particle in the ordinary thermodynamic sense. Yet we experience temperature directly. We feel warmth and cold. We associate temperature with touch. This creates the impression that temperature is a simple property sitting inside an object. But touch does not measure temperature directly. The sensation depends on the rate and direction of energy transfer between the body and the material. Metal and wood at the same room temperature may feel different because they transfer energy differently. The conscious experience is real. The physical interpretation requires more care.

The invention of thermometers transformed private sensation into public measurement. Instead of saying, "This feels hotter," observers could compare changes in the expansion of a liquid, electrical resistance, emitted radiation, or another calibrated property. Temperature became a quantity because a consistent relationship could be established between thermal condition and measurable response. The scale itself was constructed. Different temperature scales were historically possible. The physical condition existed before Celsius, Fahrenheit, or Kelvin. Human beings developed systems for representing it.

This distinction reveals the layered nature of measurement. First, a material system occupies an energetic state. Second, that state affects another system—the thermometer. Third, the instrument changes in a predictable way. Fourth, the change is mapped onto a numerical scale. Fifth, the number is interpreted through thermodynamic theory. At no point does the system contain the written value shown on the instrument. The value is a scientific expression of the state.

Temperature also shows how quantities can change meaning as theories deepen. Early ideas of heat sometimes treated it as a material fluid called caloric. This model could organize certain observations, but later developments connected heat more closely to motion, energy transfer, and microscopic behaviour. The measurable phenomena remained. The theoretical picture changed. Thermodynamics itself achieved powerful results before the detailed molecular basis of heat was fully established. Engineers could design engines and study efficiency without tracking individual particles. Again, the quantity worked before its ontology was settled.

This is a recurring pattern in physics. A measurable variable gains operational stability. It becomes useful in equations. Technology confirms its practical value. Only later does science uncover deeper layers of meaning. Temperature therefore teaches two lessons. First, a quantity can be objectively real at one scale while emerging from deeper processes at another. Second, the quantity need not exist as a substance in order to be scientifically indispensable.

This matters far beyond thermodynamics. Pressure, entropy, density, elasticity, conductivity, and even many biological variables have similar emergent character. They belong to organized systems. They describe collective manifestations. A molecule does not carry the temperature of the room. The room has a temperature because an enormous number of interactions produce a stable statistical pattern that can be measured.

From the perspective of Universal Energy Dynamics, temperature can be viewed as a manifestation variable describing the energetic distribution and exchange within a structured material form. The number does not create the thermal state. It makes the state scientifically communicable. Temperature is therefore not merely a reading. It is a vocabulary for organized energy dynamics.

4. Pressure: The Collective Consequence of Interaction

Pressure often feels more concrete than temperature. Air pushes against the walls of a container. Water presses against a dam. A sharp knife cuts more effectively because the same applied force is concentrated over a smaller area. A person feels pressure in the ears when ascending or descending. The concept seems to refer to something physically acting. Yet pressure becomes more interesting when examined across scales.

At the macroscopic level, pressure is defined as force distributed over area. This allows engineers to calculate how loads affect surfaces, how fluids behave, and how structures respond. At the microscopic level, gas pressure can be understood through the collective interactions of enormous numbers of molecules with the walls of a container. Individual collisions transfer momentum. The repeated consequences become measurable as pressure. No single molecule carries the pressure of the gas. Pressure emerges from the aggregate pattern.

This is similar to the way a crowd produces effects that no individual person produces alone. One person does not constitute crowd density. But many people distributed within an area generate a measurable collective condition. The quantity belongs to the arrangement and interaction of the system. This is why pressure can remain stable even though the individual molecules are moving unpredictably. The microscopic details fluctuate, but the macroscopic average becomes reliable. Physics succeeds by identifying that stable level.

Pressure is therefore a powerful example of selective description. A pressure gauge does not report the path of each molecule. It converts a collective effect into a single value. That single number hides enormous complexity. Millions, billions, or vastly more microscopic interactions are condensed into one measurable quantity. The compression is scientifically useful because most practical questions do not require the full microscopic history. An engineer designing a tank needs the pressure distribution, not the trajectory of every gas molecule.

The quantity is real because the collective consequence is real. A container can burst. A dam can fail. Blood pressure can affect biological function. Atmospheric pressure can alter weather and human experience. But the reality of the consequence does not require pressure to be an independent substance. It is a system-level manifestation.

The history of fluid science reflects this gradual movement from direct experience to theoretical depth. Early observers knew that fluids pushed, flowed, supported, and displaced. Later science introduced pressure as a measurable variable. Kinetic theory then connected macroscopic pressure with microscopic motion and collisions. The same quantity survived while its explanation deepened. This teaches us not to confuse persistence of vocabulary with finality of meaning. A scientific term may remain useful across several theoretical layers.

Pressure can be treated operationally in engineering, statistically in thermodynamics, molecularly in kinetic theory, and perhaps more deeply in field or substrate-based descriptions. Each layer answers a different question. What is the pressure? How does it vary? What produces it microscopically? What underlying structures permit the interaction? The answers need not compete. They form a hierarchy.

In a broader energy-dynamic view, pressure can be understood as the measurable consequence of constrained energy and momentum redistribution within or across material structures. The system is not "filled with pressure" as though pressure were a fluid. Rather, organized interactions produce a condition that pressure quantifies. This distinction may seem linguistic, but it changes how physics is understood. It shifts the focus from named entities to physical processes. Pressure is not the actor. Pressure is the measured manifestation of many interactions acting collectively through matter.

5. Charge: A Name for Interaction We Cannot See

Electric charge is more difficult to connect with ordinary experience because it is not directly visible. We see sparks, lightning, attraction, repulsion, current, chemical reactions, and electromagnetic effects. From these manifestations, physics introduces the quantity called charge. Charge is treated as a basic property of particles and systems.

It comes in positive and negative forms. It is conserved. It determines participation in electromagnetic interaction. The concept is extraordinarily successful.

Yet what is charge itself? This question is harder than calculating what charge does. A student learns that like charges repel and unlike charges attract. The statement is useful, but it risks presenting charge as a tiny hidden label attached to matter. Positive and negative may sound like intrinsic colours. But the signs are part of a relational convention. They distinguish two opposite patterns of electromagnetic response. Nature does not display plus and minus symbols on particles. The symbols organize the behaviour.

Historically, electrical phenomena were known long before charge was formulated precisely. Materials could attract small objects after rubbing. Sparks could jump. Lightning revealed immense electrical activity. Later experimenters developed systematic ways to compare and classify these effects. The quantity emerged from regularity. As with force, the phenomenon was discovered and the vocabulary was constructed.

Charge differs from mass or temperature in an important way. It is often treated as more fundamental rather than obviously emergent. Current theories assign charge a central role in field interactions and symmetries. But even here, the quantity is known through consequence. We infer charge because matter responds in consistent electromagnetic ways. We measure it through interactions, currents, fields, and transfer processes. Charge is not observed as an isolated substance separated from all behaviour. It is identified through the structure of relations.

This leaves an ontological question open. Is charge a primitive feature requiring no deeper explanation? Is it a manifestation of symmetry? Is it connected to field structure? Does it emerge from a deeper organization of a universal substrate? Physics can calculate electromagnetic effects with extraordinary precision while these deeper questions remain conceptually active. This is another example of operational success preceding final explanation.

The electron's charge is measured with remarkable consistency. The equations work. Technology built upon electromagnetism has transformed civilization. But the practical reliability of charge does not automatically tell us why such a property exists, why it has its observed magnitude, or what underlying reality makes opposite electromagnetic responses possible. The word charge can therefore conceal the mystery it names. Once we assign a symbol and unit, the concept feels complete. But naming the pattern is not explaining its origin.

This does not mean the concept is empty. It means its role should be stated accurately. Charge is a measurable manifestation variable that expresses how a physical system participates in electromagnetic interaction. The word tells us how the system behaves within the theory. It does not necessarily tell us what the system ultimately is.

In a substrate-oriented framework such as Universal Energy Dynamics, charge could be approached as a differentiated mode of structured interaction within SVE—an organized polarity or directional response expressed through matter and fields. Such a proposal would require formal development and testable distinctions. It should not be accepted simply because it sounds deeper. The disciplined question remains: what observable difference would a substrate interpretation produce? Until then, the existing quantity remains scientifically powerful.

Charge illustrates the central theme of this chapter perfectly. A physical quantity may be stable, measurable, conserved, and technologically indispensable while still functioning primarily as a vocabulary for interaction. The quantity is secure. Its deepest meaning may remain open.

6. Entropy: When a Quantity Becomes a Story

Few physical quantities have attracted as much misunderstanding as entropy. It is often described as disorder. Sometimes it is called randomness. In popular writing, it becomes decay, decline, chaos, laziness, social breakdown, ageing, or even cosmic doom. These metaphors can be suggestive, but they frequently carry entropy far beyond its precise scientific meaning.

Entropy began within the study of heat, work, and engines. Scientists wanted to understand why certain transformations occur naturally in one direction and not the reverse. Heat flows spontaneously from hotter regions to colder ones. A gas spreads through available space. Mechanical energy can dissipate into less organized forms. No ordinary engine converts all absorbed heat into useful work. Entropy emerged as a quantity that helped express these directional constraints.

Later statistical mechanics connected entropy with the number and distribution of microscopic configurations compatible with a macroscopic state. This gave the quantity a deeper interpretation: a system evolves toward conditions that can be realized in more ways. But even this statement requires care. Entropy is not an invisible force pushing systems toward disorder. It does not act like an agent. It describes the statistical and thermodynamic structure of possible configurations and transformations.

The language of disorder became popular because it offers an intuitive picture. A neatly arranged deck of cards has fewer ordered arrangements than the enormous number of mixed arrangements. A gas confined to one side of a container occupies a less statistically common macrostate than gas spread throughout the available volume. Yet "disorder" depends heavily on what is being counted and how the macroscopic state is defined. A crystal may appear highly ordered but still possess entropy. A living organism can create local structure while increasing entropy in the wider environment. A turbulent system may look disordered while containing mathematically organized patterns. The metaphor is not the quantity.

This is precisely why language matters. Once entropy is equated casually with disorder, the scientific variable becomes a universal story about everything falling apart. The story may feel profound, but it often exceeds the physics. Entropy is a powerful quantity because it identifies a deep asymmetry within physical processes. It helps explain why some transformations are available and others are overwhelmingly unlikely under ordinary conditions. But the quantity remains tied to the way states, constraints, energy distributions, and possible configurations are defined.

It is not simply a cosmic substance increasing everywhere like smoke filling a room. Entropy also reveals how difficult it can be to move between scales. At the microscopic level, many physical laws are time-reversible. At the macroscopic level, processes display a clear arrow of time. Broken glass does not spontaneously reassemble. Mixed gases do not normally separate themselves. Heat does not naturally gather from a colder body into a hotter one without additional work. The relationship between microscopic dynamics and macroscopic irreversibility has generated deep scientific and philosophical debate.

The quantity works. The meaning remains rich. In a UED-oriented interpretation, entropy could be approached as a manifestation of redistribution, accessibility, and evolving alignment within structured energy systems. Again, this would not replace thermodynamic definitions. It would attempt to place them within a broader ontology.

Entropy teaches us perhaps the strongest lesson of all: a physical quantity can begin as a precise technical concept and later become burdened by metaphysical storytelling. To preserve clarity, we must continually separate the measurement from the metaphor. Entropy does not tell the entire story of time, life, decay, or the universe. It describes one powerful aspect of how physical configurations evolve. That is already enough to make it profound.

7. Measurement as Vocabulary

Measurement is often imagined as a neutral act of reading what nature has already written. But nature does not display labels. It presents differences: longer and shorter, hotter and colder, faster and slower, more and less resistant, more probable and less probable. Measurement begins when intelligence creates a stable language for those differences.

A unit is part of that language. The metre, second, kilogram, kelvin, coulomb, pascal, and joule do not exist independently in nature as marked containers. They are agreed standards used to make observations comparable. This standardization transformed science. Without shared units, one observer's result could not be reliably compared with another's. Engineering would remain local. Experimental replication would be difficult. International science would fragment. Measurement made physical knowledge portable.

A value obtained in one laboratory could be understood elsewhere because the vocabulary had been standardized. But standardization does more than communicate results. It shapes the way questions are asked. Once a phenomenon becomes measurable, scientific attention tends to gather around the measurable variable. What cannot be easily quantified may be treated as secondary, vague, or outside physics.

This creates both power and limitation. The power is obvious. Quantities allow precise comparison. They reveal patterns invisible to casual observation. They make mathematical modelling possible. They allow technology to be designed before it is built. The limitation is subtler. Measurement always selects. It cannot capture every dimension of an event at once.

A thermometer measures thermal condition, not the human meaning of warmth. A scale measures mass, not the effort experienced while carrying the object. A pressure gauge measures collective mechanical response, not the history of each microscopic interaction. A detector records an event, not necessarily the complete ontology behind it. The unmeasured is not automatically unreal. It may be irrelevant to the current scientific question, inaccessible to the instrument, or not yet formulated as a measurable variable.

This is important because physics has often expanded by learning to measure what was previously hidden. Electricity, radiation, atomic spectra, gravitational waves, and many other phenomena became scientifically visible through new instruments and conceptual frameworks. Measurement does not merely record known reality. It helps create new windows into reality. But every window has a frame. The frame determines what appears.

This is why physical quantities can be called the manifestation vocabulary of science. They are the terms through which observable consequences become communicable, comparable, and theoretically organized. Mass tells one part of the story. Temperature tells another. Pressure, charge, and entropy reveal other relations. No single quantity contains the entire physical event. Even a complete list of quantities may still remain a model rather than the event itself.

Science therefore speaks in a structured vocabulary, not in reality's own voice. That vocabulary can become increasingly accurate. It can reveal deep regularities. It can unite phenomena once thought separate. But it should not be mistaken for the totality of existence. Measurement is the disciplined conversation between observer and manifestation. Nature provides the response. Science develops the language.

8. The Language of Science and the Risk of Reification

Science cannot proceed without language. Equations require definitions. Measurements require units. Experiments require descriptions. Theories require concepts. Even the most mathematical physics depends on words such as mass, field, state, interaction, probability, energy, symmetry, space, time, and information. These words are not decorative. They guide thought. They tell researchers what to measure, what to relate, and what questions appear meaningful.

But language carries risks. The greatest is reification—the conversion of an abstraction into an apparent thing. Once a quantity receives a noun, it begins to behave grammatically like an object. Mass resists. Charge attracts. Pressure pushes. Entropy increases. Energy flows. The sentences are efficient, but they may quietly assign agency to quantities. In reality, the system behaves. The quantity describes the behaviour.

This distinction may appear too philosophical for practical science. Engineers can calculate successfully without pausing over the ontology of every term. That is true. Operational science does not require continuous metaphysical analysis. But foundational physics does. Whenever science asks what reality is, language must be examined carefully. Otherwise, the vocabulary used for calculation becomes an untested picture of existence.

This happened historically with several concepts. Heat was once imagined as a fluid. Light was alternately pictured as particles or waves. Atoms were treated as indivisible. Space and time were regarded as absolute. The vacuum was assumed to be empty. Each concept worked within a particular framework. Each later required reinterpretation. The lesson is not that scientific language is unreliable. The lesson is that scientific language evolves.

A term can survive while its meaning changes. Energy in classical mechanics, thermodynamics, relativity, and quantum theory does not carry exactly the same conceptual picture. Mass changed under relativity. Particles changed under quantum theory. Vacuum changed under field theory. The vocabulary remains because it preserves useful continuity. But the ontology underneath it is not fixed.

This is why a quantity-centred view of physics can become narrow. It may treat the current vocabulary as though it were the final architecture of reality. A manifestation-centred view is more flexible.

It asks what observable pattern the quantity captures, under what conditions it applies, how it connects with other quantities, and what deeper processes may produce it. This does not weaken precision. It places precision within a larger structure.

For Universal Energy Dynamics, this distinction is essential. UED should not introduce SVE as merely another noun to be reified. If Structured Vacuum Energy becomes only a new invisible "thing" inserted beneath existing physics, then the conceptual advance is small. SVE must instead function as a proposal about the common structured basis through which matter, fields, gradients, and observable quantities emerge. Its value would depend on explanatory coherence and testable consequence, not on the authority of the label.

The same discipline must be applied to UED that is applied to mainstream concepts. What is observed? What is inferred? What is modelled? What is being named? What new prediction follows? Language should open inquiry, not close it. Physical quantities are among the most successful creations of scientific intelligence. They allow reality to be expressed with extraordinary precision. But they remain words, symbols, units, and relationships developed within human knowledge. The universe does not speak in kilograms, kelvins, pascals, coulombs, or joules. We do. The universe provides the stable manifestations that make such speech possible.

Closing Reflection: What the Number Conceals

Return to the heated steel railing. A thermometer now rests against it and displays a number. The number is precise. It can be recorded, compared, communicated, and inserted into equations. It can help an engineer calculate expansion, heat transfer, material stress, or safety limits. The number is useful because it reveals something stable about the physical condition. Yet the number also conceals.

It does not show every atomic interaction. It does not contain the history of sunlight absorbed by the metal. It does not contain the child's sudden pain. It does not explain why matter can store and redistribute energy. It does not reveal the deepest structure that allows thermal behaviour to exist. It selects one manifestation and expresses it in scientific language.

The same is true of mass. The number on the scale captures a stable relation, but it does not reveal the complete architecture of matter. Pressure condenses countless microscopic interactions into one value. Charge summarizes a consistent pattern of electromagnetic response. Entropy expresses the statistical and thermodynamic structure of evolving configurations. Each quantity tells the truth. But each tells a selected truth.

This is not a weakness of science. It is the source of its power. Science succeeds because it does not attempt to describe everything at once. It isolates stable relations, defines measurable variables, and builds theories from them. The danger begins only when the selected truth is mistaken for the complete reality.

Physical quantities are not meaningless inventions. They are disciplined constructions constrained by the behaviour of nature. They are not arbitrary. But neither are they necessarily the fundamental furniture of existence. They are the manifestation vocabulary through which science converts physical consequence into communicable knowledge.

This gives us a clearer hierarchy:

- Matter and fields behave.
- Energy redistributes.
- Systems interact.
- Manifestations appear.
- Instruments respond.
- Quantities are assigned.
- Equations relate them.
- Theories organize them.

- Interpretations attempt to explain what lies beneath.

The quantity belongs to the middle of this hierarchy. It is neither the raw event nor the final ontology. It is the bridge. Once this is understood, physics becomes broader. Mass is no longer simply a substance. Temperature is no longer merely a number. Pressure is no longer an invisible thing pushing. Charge is no longer a mysterious sign attached to a particle. Entropy is no longer a universal myth of disorder. Each becomes a precise way of describing how a structured physical system manifests under particular conditions.

And beneath every quantity, the deeper question remains: *What is reality doing that makes this measurement possible?* That question does not diminish the language of science. It gives the language its proper place. A number may describe reality with remarkable accuracy. But reality is always more than the number.

Interlude I

When Quantities Begin to Disappear

By the end of the first three chapters, an important shift has already occurred. Force no longer appears as a simple thing travelling between objects. Nature no longer appears to contain textbook categories exactly as human beings name them. Mass, temperature, pressure, charge, and entropy no longer look like labels written inside matter.

They remain measurable. They remain useful. They remain scientifically real.

But they begin to change position. Instead of appearing as the basic furniture of the universe, they begin to appear as the vocabulary through which science describes recurring physical manifestations.

This is not the disappearance of quantities from physics. It is the disappearance of their apparent finality.

From Experience to Measurement

A quantity often enters science when experience becomes too vague. A person can say that one stone feels heavier than another, but physics asks *how much heavier*. A person can say that water is hot, but physics asks *what temperature has been measured*. A person can feel resistance, but physics asks *what force, pressure, or deformation is involved*.

Measurement takes a private or qualitative experience and converts it into a shared relation. That conversion is one of the great strengths of science. It allows comparison. It allows repetition. It allows prediction.

It allows knowledge to leave the body of one observer and become available to many.

But every successful conversion creates a temptation. Once the quantity is precise, we may stop asking what process has been compressed into it.

- A force value may hide deformation, field response, momentum transfer, surface interaction, internal stress, and surrounding constraints.
- A pressure reading may hide the collective consequences of countless microscopic collisions.
- A temperature value may hide a distribution of energetic states.
- A mass value may hide questions about inertia, internal energy, fields, and material organization.
- A charge value may hide a deeper pattern of electromagnetic participation.

The number becomes clear while the process becomes invisible.

When Quantities Require Explanation

Scientific progress frequently begins when someone notices what the quantity is concealing.

- A thermometer measures temperature accurately, but the measurement alone does not explain what heat is.
- A pressure gauge records a value, but the value does not reveal every microscopic interaction producing it.
- A balance compares mass, but the reading does not settle the deepest meaning of inertia.
- A force sensor records mechanical response, but it does not tell us whether force is fundamental or whether it emerges

from a deeper physical structure.

The quantity answers one question. Its success exposes another. Physics therefore moves through a recurring sequence:

1. A phenomenon is experienced.
2. It is named.
3. It is measured.
4. The measurement is connected with other quantities.
5. The relation is expressed mathematically.
6. The quantity that once appeared explanatory begins to require explanation itself.

Heat was once treated almost like a substance. Later, thermodynamics and statistical mechanics revealed deeper relations among energy, motion, distribution, and collective behaviour. Gravity was described as force. Later, general relativity described free motion through spacetime geometry. Particles appeared to be tiny separate objects. Later, quantum field theory connected particle-like behaviour with field excitations.

At each stage, the earlier quantity did not become useless. It became located within a wider structure.

The Disappearance of Finality

This is how quantities begin to disappear. They do not vanish from calculation. They disappear from the deepest layer of explanation.

Temperature remains indispensable, but no one expects to find a small object called temperature inside a gas. Pressure remains measurable, but pressure is not a separate material pushing from within. Force remains central to mechanics, but gravity no longer needs to be treated as an ordinary pulling agent in the relativistic description.

The quantity survives operationally while its ontology changes.

This movement is easy to miss because textbooks usually present the result after the conceptual struggle has ended. A student learns the quantity, its unit, and its equation. The quantity appears complete because the deeper problem is not visible.

Mass is measured in kilograms. Force in newtons. Pressure in pascals. Temperature in kelvin or degrees. The categories seem naturally present in the universe. But scientific history tells a different story.

Each quantity emerged through attempts to isolate one stable relation from a more complicated event. Each became useful because nature constrained it. Each gained authority because repeated observations supported it. And each remains open to reinterpretation when physics moves into a deeper descriptive layer.

Quantities as Bridges

The lesson is not that quantities are false. The lesson is that their reality is relational.

A metre is not hidden inside a road. A second is not stored inside an event. A newton is not a substance inside a collision.

The units are created. The differences they measure are not. Nature produces structure, change, resistance, motion, distribution, and consequence. Human intelligence develops quantities to represent those recurring differences.

This places the quantity between reality and theory. It is neither pure invention nor independent substance. It is a disciplined bridge. But a bridge should not be mistaken for the landscape it crosses.

From Objects to Surroundings

Once this distinction becomes visible, physics begins moving away from isolated objects carrying fixed properties. It begins looking toward the region around them.

- What conditions surround the object?

- What changes across space?
- What pattern extends beyond the visible body?
- What relation connects distant events?
- What allows one system to respond to another without direct mechanical contact?

The quantity points beyond itself.

Force points toward field. Potential points toward spatial structure. Charge points toward electromagnetic relation. Mass points toward inertia, energy, and gravity. Temperature points toward collective microscopic states. Pressure points toward distributed interaction. Entropy points toward configuration, probability, and accessibility.

The measurement does not close the event. It becomes an entrance into a deeper level.

This is where the object-centred picture begins to weaken. In ordinary experience, reality appears to consist of separate things. A stone is here. Another stone is there. A magnet is here. A piece of iron is there. Earth is here. The Moon is far away.

If one object affects another, the first question seems obvious: *What force passed between them?*

But once force is treated as a measured response rather than a travelling object, another question appears: *What physical condition exists between and around the bodies?*

The space that once appeared empty begins to acquire significance. The region around the magnet cannot remain irrelevant if the iron responds differently at different positions. The region around a charge cannot remain featureless if another charge responds according to location. The region around Earth cannot remain merely empty distance if motion changes systematically throughout it.

When Quantities Begin to Disappear

The search for deeper explanation moves outward from the object. Physics begins to investigate distributed structure.

This transition required imagination. The visible object was easy to accept. The invisible field was harder. A field could not be held like a stone. It could not always be seen directly. It was known through consequences.

A compass needle changed direction. A charge accelerated. Light propagated. A planet followed a path.

The field became physically meaningful because its structure organized observable responses. This was an important scientific movement: what cannot be directly seen may still become real through consistent, measurable consequence.

But the same caution must remain. A field is not explained merely by naming it. Calling an interaction *field-mediated* is deeper than saying one object mysteriously acts upon another, but it does not necessarily settle what the field fundamentally is. The concept relocates the question. It gives physics a more powerful structure, but the search continues.

The Recurring Pattern

The history of physics repeatedly follows this pattern:

- A quantity solves one problem.
- Its success reveals a hidden assumption.
- A new concept enters.
- The new concept unifies several quantities.
- Then the new concept itself becomes the subject of deeper inquiry.

Force leads toward fields. Fields lead toward geometry, quantum states, and vacuum structure. Matter leads toward atoms. Atoms lead toward particles. Particles lead toward quantum fields. Empty space leads toward a vacuum with physical properties.

At no stage does the earlier description become meaningless. It becomes part of a larger architecture.

Standing at the Transition

The reader is now standing at that transition. The first part of this book began with experience. It asked how push, pull, resistance, motion, heat, and pressure became scientific quantities. It separated nature from the vocabulary used to describe it. It examined the hidden meanings inside familiar measurements.

Now the direction changes. The question is no longer only: *What does the quantity describe?* It becomes: *What physical structure makes the quantity possible?*

The answer cannot be found by looking only at isolated objects. We must begin looking between them, around them, and perhaps beneath them.

Seeing Through the Quantity

The quantity remains on the page. The instrument still records it. The engineer still uses it. The student still calculates it. But in the search for deeper reality, the quantity begins to become transparent. We start seeing through it.

- Behind force, there may be a field.
- Behind pressure, collective interaction.
- Behind temperature, organized distribution.
- Behind mass, stable physical structure.
- Behind the object, a surrounding condition.
- Behind the visible event, an invisible architecture.

The next stage of the journey therefore begins where apparently empty space stops being empty. Physics turns from objects toward fields. And the region between things becomes part of reality.

Chapter 4 — From Objects to Fields

The Empty Space Between Things

Place two magnets on a table. Move them slowly toward one another. Before they touch, something happens. One magnet may pull the other closer. Turn one around, and the interaction changes. Now they resist. The space between them appears empty, yet the behaviour of each magnet depends on the presence and orientation of the other.

Nothing visible connects them. No rope stretches between them. No hand pushes from the middle. Still, motion changes.

For most of human history, physical reality was imagined mainly through objects. Stones, planets, bodies, machines, and particles appeared to be the real things. Empty space merely separated them. Objects acted. Objects collided. Objects pushed and pulled. Space was the stage. Matter was the actor.

This picture worked naturally because human experience is object-centred. We see surfaces, boundaries, shapes, and motion. We handle tools, lift weights, strike materials, and observe collisions. The most obvious physical events occur where visible things meet. But many important phenomena did not fit comfortably inside this picture.

Gravity acted across distance. Magnets influenced one another without contact. Electric charges attracted and repelled. Light travelled through regions that appeared empty. The old question returned repeatedly: how can one object influence another when nothing visible passes between them?

Science first answered through force. Then it moved toward fields. Later, gravity became geometry. At the quantum level, particles themselves began to appear less like permanent miniature objects and more like localized manifestations of underlying fields. The conceptual centre of physics gradually shifted.

The universe no longer looked like a collection of isolated things moving through emptiness. It began to look like a continuous structure in which local conditions, fields, geometries, and interactions shaped what objects could do. Force did not vanish from calculation. Engineers still calculate forces. Students still draw force diagrams. Machines still succeed or fail according to mechanical loads. But force slowly lost its status as the deepest explanation.

What appeared to be a direct push became a response to a field. What appeared to be gravitational attraction became motion through curved spacetime. What appeared to be a particle became an excitation of a quantum field. This history reveals something important about physics: science does not always progress by adding more objects. Often it progresses by dissolving objects into deeper relations. The visible thing remains. The explanation moves beneath it.

1. When Physics Was a World of Objects

Ordinary perception divides reality into things. A stone is one thing. A tree is another. The ground lies beneath them. The air fills the space between them. This object-based view is not foolish. It is highly effective at the scale of daily life. Solid bodies retain shape, occupy regions, collide, break, rotate, and move. Much of classical mechanics can be developed by treating them as distinct entities with measurable properties.

Early physical thinking therefore focused naturally on objects and their motions. The central questions were simple to state: why does a body move? Why does it stop? What keeps planets in orbit? What happens when bodies collide? How does one object produce change in another?

The answer usually took the form of direct action. A hand pushes a cart. A hammer strikes a nail. A rope pulls a load. A surface resists motion. In each case, visible contact appears to produce a visible consequence. This reinforced the idea that causation required touching. A moving object transferred motion to another. A body changed only when something acted directly upon it.

But contact itself becomes less simple when examined closely. When two surfaces appear to touch, their microscopic structures interact through electromagnetic effects. The atoms do not merge into one another. The sensation of solidity arises from resistance within organized matter. Even the most ordinary collision already contains deeper field behaviour.

Classical physics did not need this microscopic explanation in order to calculate motion. It treated bodies as solid units because that approximation worked. This is one of the strengths of science: it can describe one level accurately without knowing every deeper level.

Yet the object-centred picture faced greater difficulty with non-contact interaction. A falling stone moved toward Earth without a visible connector. The Moon remained associated with Earth across vast distance. A magnet influenced iron without touching it. Electrified materials attracted small objects through empty space. These effects seemed to require action where no direct contact could be seen.

One possibility was that invisible matter filled space and transmitted influence. Another was that bodies could somehow act directly across distance. Both possibilities were conceptually uncomfortable. Invisible media could easily become speculative if they produced no independent measurable consequences. Action at a distance seemed almost magical because it allowed one object to affect another without an intervening mechanism.

The object-based worldview had reached a boundary. It could describe the bodies clearly. It struggled to explain the space between them. This is where the idea of force became essential. Force allowed physics to quantify interaction without first resolving the complete mechanism. If an object accelerated in a measurable way, a force could be assigned. The concept organized behaviour. But it did not necessarily explain what occupied the gap.

This distinction remained hidden because calculations worked so well. Once force entered equations, the empty space between bodies appeared mathematically manageable. Yet the deeper question had not disappeared. What carries the interaction? Does space participate? Is the influence local or instantaneous? Is force a real entity, or a label for the observed response?

These questions eventually pushed physics away from isolated objects and toward a more distributed view of reality. The world of objects had not become false. It had become incomplete.

2. Newton and the Problem of Action at a Distance

Newtonian mechanics unified terrestrial and celestial motion with extraordinary power. The same principles could describe falling objects, projectiles, tides, the Moon, and planetary orbits. This was a major intellectual shift: Earth and heaven no longer required separate kinds of physics.

But Newton's gravitational theory carried a conceptual difficulty. Two masses could influence each other across empty space. The mathematical relation was clear. The underlying mechanism was not. How could the Sun affect Earth without touching it? How could Earth affect the Moon? Did gravitational influence travel? Was it instantaneous? Did space contain something capable of transmitting it?

Newton was cautious about inventing a hidden mechanism without evidence. His restraint is important because it shows that even the creator of a highly successful theory distinguished mathematical description from ultimate explanation. The equations worked. The cause beneath them remained unsettled.

This is a recurring pattern in science. A theory may reach operational maturity before achieving ontological clarity. Newtonian gravity provided reliable predictions. Astronomers could calculate orbits. Engineers could study motion. The theory became foundational. Yet action at a distance continued to trouble physicists and philosophers.

The difficulty was not merely emotional. It concerned the structure of causation. In ordinary contact interaction, change appears to propagate through material continuity. A hand pushes a rod, the rod transmits stress, and another object moves. There is an identifiable chain. In gravitational action at a distance, the chain seemed absent. The source mass was here. The responding mass was there. The influence appeared in between without a visible carrier.

One way to manage this was to treat force purely as a mathematical relation. It did not matter what force was, provided the equation predicted correctly. This operational attitude was enormously productive. But it also left the foundational question untouched. Another possibility was to fill space with a medium. Historically, physicists often proposed forms of ether to support light, electrical action, or gravity. Such proposals arose naturally because known waves seemed to require media. Sound travelled through air. Water waves moved through water. Perhaps light and other influences travelled through something subtler.

The problem was that a medium must do more than satisfy imagination. It must generate measurable consequences. When no clear evidence distinguished one mechanical picture from another, the proposed medium risked becoming explanatory decoration. This is where physics faced an important discipline: do not invent hidden machinery merely because the mathematics feels conceptually incomplete. But the opposite danger also existed: do not confuse successful equations with a complete account of reality.

Newtonian force occupied the space between these two cautions. It allowed physics to proceed without pretending that the mechanism was settled. The result was a model of extraordinary range, but one in which force remained somewhat mysterious. This mystery became less tolerable as electrical and magnetic phenomena developed. Their patterns were more complex than simple attraction along a line between two bodies. Orientation, induction, changing currents, and wave behaviour suggested that the surrounding region itself had physical structure. Space could no longer remain a passive gap. The problem of action at a distance was preparing physics for the field.

3. Faraday: Seeing Structure Where Others Saw Space

Michael Faraday changed physics partly because he thought differently. He did not begin with advanced mathematical formalism. He began with experiments, patterns, and physical intuition. Where others focused mainly on interacting bodies, Faraday paid attention to the region around them.

Iron filings placed near a magnet arranged themselves into visible curves. The pattern suggested that magnetic influence was not merely a direct pull between isolated objects. The surrounding space appeared structured. Faraday developed the idea of lines of force. These lines were not simply decorative drawings. They represented his conviction that electrical and magnetic influence existed throughout space and could be understood locally.

This was a conceptual revolution. Instead of saying that one object acted mysteriously on another from afar, field-like thinking said that the source altered the physical condition of the surrounding region. Another object then responded to the condition where it was located. Influence became local. The intervening space gained physical importance.

This did not immediately settle what the lines themselves were. Were they literal structures? Visual aids? Tensions in an unseen medium? Mathematical representations? Faraday's language carried both physical intuition and conceptual uncertainty. But the shift was decisive. The space between objects was no longer empty in the explanatory sense. It possessed organization.

Faraday's work on electromagnetic induction made this especially clear. A changing magnetic condition could generate an electrical effect. The relation depended not merely on static objects but on changing structures distributed through space. The old image of fixed bodies exerting simple forces became increasingly inadequate.

Fields could vary. Fields could store energy. Fields could propagate changes. Fields could interact with matter locally. The focus moved from what objects did to one another toward how the surrounding physical condition evolved.

Faraday's intellectual position was not immediately dominant. His visual and experimental style differed from the mathematical habits of many established physicists. Lines of force could appear too intuitive or insufficiently formal. Yet this tension between physical intuition and mathematical authority became productive. Faraday saw the structure. Maxwell later gave the structure mathematical form.

This episode reveals how science often advances through complementary modes of intelligence. One scientist notices a physical pattern that existing mathematics does not yet express. Another develops the formal system capable of organizing it. The eventual theory appears clean in textbooks, but its origin lies in conceptual struggle.

Faraday's field idea also demonstrates that scientific progress sometimes requires granting reality to something that cannot be directly seen. No one saw the magnetic field itself. They saw its consequences. Iron filings aligned. Currents changed. Needles turned.

Forces appeared. The field was inferred as the organized physical condition connecting these manifestations.

This raises an important philosophical question. Is a field discovered or invented? The observable regularities are discovered. The field concept is constructed to represent and unify them. But as the concept predicts new phenomena and supports independent measurements, it gains scientific reality. The field becomes more than a metaphor. Still, its deepest ontology remains open.

Faraday's contribution was therefore not that he found a visible substance filling space. It was that he changed the unit of physical thinking. The basic reality no longer had to be the isolated object. It could be the structured relation extending through space. That move prepared physics for Maxwell—and for the gradual disappearance of force as the ultimate explanatory category.

4. Maxwell and the Rise of the Field

James Clerk Maxwell transformed Faraday's physical intuition into a mathematical theory. This was one of the decisive moments in the history of physics. Electricity and magnetism had often been studied as related but distinct phenomena. Maxwell's equations revealed them as aspects of a unified electromagnetic field.

Even more remarkably, the equations implied waves travelling through space at a characteristic speed. That speed matched the known speed of light. Light itself could therefore be understood as an electromagnetic wave. Three domains—electricity, magnetism, and optics—were drawn into one structure.

This is what deep theories often do. They do not merely improve individual calculations. They reveal that phenomena previously treated as separate are manifestations of a common underlying pattern.

The field now became mathematically dynamic. It was not merely a static region around a charge or magnet. Changing electric fields were connected with magnetic fields. Changing magnetic fields were connected with electric fields. The disturbance could propagate. The field possessed its own behaviour.

This weakened the older idea that physical reality consisted only of matter acting on matter. An electromagnetic wave could travel through space even after leaving its source. The field could carry energy and momentum. The field no longer looked like a convenient bookkeeping device attached to objects. It appeared to possess physical significance of its own.

Yet this success introduced another dilemma. If light was a wave, what was waving? Known mechanical waves required media. Sound required matter. Water waves required water. Many physicists therefore imagined an electromagnetic ether filling space. The ether seemed necessary because the field was still interpreted through mechanical analogy. If the wave existed, something must be oscillating.

This assumption appeared reasonable. It also shows how older concepts shape new theories. Even after physics discovered the field, it tried to understand the field as motion within a hidden substance resembling familiar matter. The electromagnetic theory worked. The mechanical picture of the supporting ether remained speculative.

Experiments searching for Earth's motion through the ether failed to produce the expected result. The Michelson–Morley experiment became especially famous in this context. Its null result did not instantly settle every debate, but it exposed a serious problem for simple ether-drift expectations. Physics now faced a difficult choice. Should it preserve the imagined medium by adding more assumptions? Or should it rethink space, time, and motion themselves?

Einstein's special relativity later removed the need for the classical mechanical ether in electromagnetic theory. The speed of light became part of a broader structure of spacetime rather than the speed of a wave through a material medium.

This historical episode is highly instructive. The field theory was successful. One interpretation of the field—as vibration in a mechanical ether—was not required by that success. The mathematics captured real regularities. The accompanying picture was provisional.

Maxwell's work therefore represents both a triumph and a warning. It showed that distributed fields could be more fundamental than direct forces. But it also showed how easily scientists can attach an unnecessary ontology to a successful formalism. The field itself solved one layer of the problem. The question of what the field fundamentally was remained.

Still, the conceptual direction was irreversible. Space could carry structured physical activity. Interactions could propagate locally. Force increasingly appeared as the measurable response of matter to a field already present around it. Instead of asking, "What force reaches across the gap?" physics could ask, "What is the field at this location, and how does matter respond to it?" The object remained visible. The field became explanatory.

5. Einstein: When Gravity Stopped Being a Force

Einstein pushed the field concept into an even deeper transformation. Special relativity altered the relation between space, time, motion, energy, and momentum. General relativity then reinterpreted gravity itself.

In Newtonian mechanics, gravity is a force of attraction between masses. In general relativity, freely falling bodies are not necessarily being pulled in the classical sense. They follow natural paths through curved spacetime. Matter and energy shape spacetime geometry. Geometry guides motion.

This changed the conceptual status of gravity. A falling object can be described as accelerating under gravitational force in the Newtonian framework. In general relativity, the same object may be treated as moving freely along a geodesic within curved spacetime. The observable event remains. The explanation changes.

This is one of the clearest examples of force disappearing into a deeper structure. The transformation began from a simple but profound observation: locally, gravitational acceleration and acceleration of a reference frame can appear equivalent. A person inside a sealed accelerating lift may feel pressed against the floor. A person standing on Earth also feels weight. Conversely, a freely falling person may temporarily feel weightless. These experiences suggest that gravity is not merely an ordinary force acting within fixed space and time. The structure of space and time themselves must participate.

General relativity gave this insight mathematical form. The theory explained the small anomaly in Mercury's orbit, predicted the bending of light near massive bodies, and later supported further observable consequences. It succeeded because it changed the framework rather than merely adjusting the force law.

This is important. Scientific progress does not always come from adding another force, particle, or correction. Sometimes it comes from changing the geometry of the problem. What looked like an interaction between objects becomes a feature of the structure through which they move.

Yet even general relativity does not eliminate all conceptual questions. What is spacetime physically? Is curvature a literal physical structure, a geometrical description of relations, or both? Does spacetime emerge from something deeper? Can it be reconciled completely with quantum theory? The theory is extraordinarily successful. Its ontology remains open to interpretation and further development.

Einstein himself worked within a scientific culture still struggling with the boundary between field and matter. In his mature thinking, the field acquired great importance. The old picture of particles moving through an independent background became increasingly unsatisfactory. But relativity also complicated the field idea. The gravitational field was no longer simply something located within spacetime. It was connected with spacetime structure itself. The stage had become part of the action.

This erased another division inherited from ordinary experience. Previously, objects existed in space and changed with time. Now space and time formed a dynamic physical geometry influenced by matter and energy. Physics had moved from objects, to forces, to fields, to geometry.

At each stage, the previous description remained useful within appropriate conditions. Newtonian gravity did not become useless. It remained an excellent approximation for many practical situations. But its ontological status changed. What classical mechanics called force could now be understood as a manifestation of deeper geometric structure.

This is the pattern the present book keeps emphasizing: a quantity remains operationally valid after losing its claim to fundamentality. Force survives in engineering. Gravity survives in ordinary speech. Weight survives in bodily experience. But beneath them, the theory moves deeper. The disappearance of force is therefore not an erasure. It is a relocation. Force moves from the foundation of explanation to the level of observable consequence.

6. Quantum Fields and the Fading Particle

The quantum revolution disrupted the object-centred picture even more radically. Classical particles resemble tiny bodies. They possess positions, trajectories, and stable identities. Even when invisible, they are imagined as miniature versions of ordinary objects.

Quantum experiments made this picture difficult to sustain. Light displayed both wave-like and particle-like behaviour. Matter also produced interference. Particles did not always follow definite classical paths. Measurement outcomes were represented probabilistically. The distinction between object and process became unstable.

Quantum mechanics provided an extraordinarily successful mathematical framework, but its meaning generated deep disagreement. Were particles real objects with hidden properties? Was the wavefunction physically real? Did measurement create definite outcomes? Was the theory incomplete? Did reality branch? Were quantum properties relational rather than absolute? Physicists could often agree on calculations while disagreeing over ontology.

Quantum field theory shifted the conceptual centre again. In this framework, fields are fundamental to the formalism, and what we call particles are treated as excitations or quanta of those fields. An electron is not simply a tiny solid bead travelling through emptiness. It is associated with the electron field. A photon is an excitation of the electromagnetic field. Other particle types correspond to other quantum fields. The particle becomes a localized manifestation of a distributed underlying structure.

This resembles the earlier movement from force to field, but now the object itself begins to dissolve. The field is not merely something surrounding the particle. The particle arises from the field. This is a major conceptual inversion.

In classical thinking: objects generate fields. In quantum field thinking: fields can generate what appear as particles.

The vacuum also changes meaning. It is no longer simple emptiness. Quantum fields remain present even in states with no ordinary particles. Vacuum fluctuations, zero-point structures, and measurable effects associated with field states make the idea of absolute nothingness increasingly difficult to maintain.

Yet caution is necessary. Quantum fields are not directly seen as material substances. They are elements of an immensely successful mathematical and experimental framework. Their scientific reality is supported by predictive success, consistency, and observed consequences. But what quantum fields ultimately are remains philosophically difficult.

Are they physical entities spread through space? Are they mathematical structures? Are they excitations of something deeper? Does spacetime itself emerge from a more fundamental quantum structure? The formalism works. The ontology remains debated.

This is precisely why quantum field theory should not be used carelessly as final proof that "everything is fields." Such a statement may express a useful contemporary viewpoint, but it risks turning a powerful theoretical language into an unquestioned metaphysics.

Still, the historical direction is clear. The independent particle has lost its classical solidity. The vacuum has lost its emptiness. The field has gained explanatory centrality. Force has become even more secondary. Interactions are described through field couplings, symmetries, exchange processes, and changes of state.

What appears macroscopically as attraction, repulsion, pressure, radiation, or impact emerges from deeper field dynamics. This does not mean that everyday objects are unreal. A table remains stable enough to support weight. A particle detector records localized events. Matter forms persistent structures. But the deeper explanation no longer resembles a universe made from tiny hard pieces.

The object is real at one scale. The field may be more fundamental at another. Physics continues moving downward—not necessarily toward smaller things, but toward deeper relations.

7. Why Physics Keeps Moving Toward Deeper Layers

The repeated movement from objects to forces, from forces to fields, from fields to geometry, and from particles to quantum fields is not accidental. Science moves deeper because every successful description eventually exposes its own boundary.

Objects explain visible collisions but not distant influence. Force quantifies interaction but may not reveal the mechanism. Fields localize interaction but raise questions about what fields are. Geometry explains gravity but leaves unresolved the relation between spacetime and quantum theory. Quantum fields organize particles and interactions but still leave foundational and ontological questions open.

Each layer solves a problem and creates a new frontier. This is not evidence that physics is failing. It is evidence that reality is deeper than any single vocabulary. Science keeps moving because observations resist simplification. Small discrepancies matter. Unexpected results matter. Conceptual contradictions matter.

A theory may work so well that its limitations remain hidden for generations. Then a new measurement reveals a deviation. An old assumption becomes visible. A deeper framework becomes necessary. Mercury's orbit challenged the completeness of Newtonian gravity. Electromagnetic theory challenged absolute space and time. Quantum experiments challenged classical objects and trajectories. The expansion of cosmology challenged simple pictures of matter and vacuum.

Progress often begins where a successful theory almost works. This "almost" is scientifically precious. It shows where the map stops matching the territory. Institutions naturally favour established frameworks because science requires shared standards. Laboratories, textbooks, journals, and funding systems cannot rebuild their foundations every week. Stability is necessary. But too much stability can turn an effective theory into an unquestioned worldview.

A deeper layer usually appears first as an inconvenience. It may look unnecessary, speculative, or philosophically troublesome. Only after it produces clearer explanation or distinctive prediction does it become accepted. The lesson is not that every new proposal deserves equal status. Most speculative ideas will not survive. The lesson is that established success should not be mistaken for finality.

This is where the UED doorway can be opened carefully. If modern physics has already moved from objects toward fields, and if fields themselves remain conceptually unresolved, then asking whether fields arise from a deeper universal substrate is scientifically legitimate. The question is not absurd. The answer, however, cannot be accepted merely because it sounds coherent.

Universal Energy Dynamics proposes Structured Vacuum Energy, or SVE, as the underlying continuous basis through which matter, fields, and physical manifestations emerge. In this view, matter is not placed inside an empty background. Matter is an organized form of the substrate. Fields are not separate additions surrounding matter. They are structured modes of substrate behaviour. Forces are not independent agents. They are measurable consequences of local gradients, alignments, and interactions within that structure.

This picture follows the historical direction of physics while extending it. But extension is not proof. For SVE to become more than a philosophical proposal, it must produce clear formal relations, testable distinctions, and explanatory gains beyond existing theories. The discipline remains the same. What does the model predict? What observation could support or challenge it? Where is it interpretation? Where is it measurable physics?

The goal is not to replace "field" with "substrate" as another unquestioned noun. The goal is to investigate whether a deeper common structure can explain why fields, matter, and quantities appear at all. Physics keeps moving toward deeper layers because descriptions are not reality. They are successive openings into reality.

Closing Reflection: When the Space Became Physical

Return to the two magnets. They remain separated on the table. No visible object joins them. Yet the behaviour of one depends on the other. At first, we may describe this as a force acting across space. Then we speak of a magnetic field. At a deeper level, we describe the field through equations, electromagnetic structure, and quantum interactions. The observable event has not changed. Our understanding has.

This is the history of physics in miniature. Human beings first noticed objects. Then they noticed interactions. Then they quantified those interactions as forces. Then they discovered that the region around objects could possess structure. Fields replaced empty space. Geometry replaced gravitational force. Quantum fields weakened the idea of permanent particles. At each stage, something that once appeared fundamental became a manifestation of something deeper.

The object did not disappear. The force did not become useless. The field did not make engineering obsolete. Each concept retained value within its domain. But the hierarchy changed. Force became less like a cause and more like a measurable consequence. The field became less like an invisible tool and more like a central physical structure. The particle became less like a tiny object and more like a localized field excitation. Space became less like emptiness and more like an active part of physical reality.

This progression should make us both confident and humble. Confident, because physics has repeatedly discovered deeper coherence beneath apparently separate phenomena. Humble, because every deeper layer has eventually produced new questions. What is a field? What is spacetime? What is the vacuum? Why do quantum fields exist? How do matter and geometry emerge? Are these final structures—or descriptions of an underlying substrate still beyond present theory?

The disappearance of force therefore does not end the inquiry. It sharpens it. We no longer need to ask only: *What force acts between these objects?* We can ask: *What field structure produces this response? What geometry shapes this motion? What deeper dynamics give rise to the field?* And perhaps, eventually: *What universal foundation allows objects, fields, space, and energy to appear as different manifestations of one reality?*

The two magnets still sit on the table. The space between them no longer looks empty. That change in vision may be one of the most important achievements in the history of physics.

Chapter 5 — Is Force Really Fundamental?

What Do You Actually Feel?

Stand quietly on the floor. Nothing appears to be happening. You are not moving. The room is still. Yet according to ordinary physics, Earth is exerting a gravitational force on your body. You call this force weight.

Now imagine that the floor suddenly disappears. You begin to fall. Gravity is still present. Your body accelerates toward Earth. Yet during free fall, the familiar sensation of weight begins to disappear. If you and everything around you fall together, you may feel temporarily weightless.

This creates a strange situation. When gravity is supposedly acting freely, you may not feel it. When you stand motionless on the ground, you do. What, then, are you actually feeling?

You are not directly feeling gravity pulling every part of your body downward. You are feeling the floor preventing your body from following its natural falling path. The floor pushes upward against your feet. That supporting interaction is transmitted through bones, muscles, joints, and tissues. Your body compresses slightly under the constraint. The sensation called weight arises largely from resistance to free fall.

This distinction is easy to miss because everyday language combines several layers. We say Earth pulls us downward, the ground holds us up, and our body feels heavy. These statements are useful at the ordinary scale. But they do not all refer to the same physical process.

A person standing on the floor feels support. A person falling freely feels little support. An astronaut orbiting Earth is still strongly influenced by gravity, yet appears weightless because the spacecraft and astronaut are falling together. Gravity has not vanished. The experience of force has changed.

This suggests that force may not be a simple physical substance delivered to an object. What we call force may sometimes describe the relation between a system, its environment, its constraints, and its changing motion.

A body does not need to "feel" force in the conscious sense. It responds. A spring deforms. A planet changes trajectory. A charged particle accelerates. A material surface resists penetration. A fluid flows from one condition to another. We observe these responses and describe them through force.

This description is enormously useful. But usefulness alone does not establish fundamentality. A quantity may accurately capture a consequence while remaining secondary to the structure that produces it. Temperature describes a collective energetic condition without being a microscopic substance. Pressure describes distributed interaction without being an independent material. Force may occupy a similar position.

Perhaps objects do not respond because force exists as a separate thing. Perhaps force is the name given to their measurable response within fields, gradients, geometries, and changing configurations. The question, then, is not whether force is real enough to calculate. It clearly is. The deeper question is: *Is force part of reality's foundation, or is it the visible vocabulary of something deeper?*

1. Why Objects Do Not "Feel" Force

The statement that objects do not feel force requires care. A living person can consciously experience pressure, pain, strain, balance, and acceleration. A non-living object cannot report experience. Yet even in a physical sense, an object does not detect a labelled quantity called force. It undergoes changes within its structure and motion.

A metal rod under tension stretches slightly. A rubber band changes shape. A ball struck by a bat accelerates and deforms. A bridge under load develops internal stresses. A planet follows a curved path. The object responds according to its material structure, inertia, surrounding conditions, and internal interactions. The label force belongs to the scientific description of that response.

This becomes clearer when we distinguish between whole-body motion and internal deformation. Suppose a spacecraft engine fires in deep space. The engine interacts with expelled material, and the spacecraft accelerates. People inside feel pressed against their seats because the seats accelerate their bodies. The contact force is transmitted unevenly through the body. The back is pushed first; internal tissues then respond. The sensation arises because different parts of the body are being constrained and accelerated through material contact.

Now compare this with free fall in a gravitational field. Gravity influences the body throughout its structure. If the field is approximately uniform across the body, all parts fall together. There may be little internal compression, so the person feels weightless. The body is accelerating relative to Earth, but it does not experience the same internal stress as it does while standing on the ground.

This distinction between coordinate acceleration and felt acceleration became central to Einstein's reasoning about gravity. What a person physically feels is often connected not to motion alone, but to resistance, constraint, deformation, and what relativity calls *proper acceleration*—the acceleration registered by an accelerometer moving with the person.

An accelerometer resting on the floor records an upward proper acceleration because the floor prevents free fall. An accelerometer falling freely ideally records zero, even though its distance from Earth changes rapidly. This overturns ordinary intuition. The standing person feels a force-like effect because natural motion is being interrupted. The falling person feels little because the body is following the local gravitational structure freely.

The same principle appears in less dramatic situations. When a car turns, passengers feel pushed sideways. Yet the important physical interaction is usually the seat, door, or friction redirecting the body. The body tends to continue along its previous path while the vehicle changes direction. When a lift accelerates upward, a passenger feels heavier because the floor must produce a greater supporting interaction. When the lift accelerates downward, the person feels lighter because less support is required. The felt experience varies even though the person's mass remains unchanged.

Force is therefore not a single sensation corresponding directly to one external entity. The experience is the biological reading of material responses. This is why phrases such as "the object feels a force" must be understood metaphorically. In mechanics, we mean that the object's motion or structure changes in a manner described by a force relation. The object does not identify a cause. It participates in a process.

The deeper physical reality may consist of field conditions, energy distributions, constraints, momentum exchange, geometric structure, and internal material response. Force gathers these into a convenient measurable category. It is a highly successful summary. But a summary should not automatically be treated as the underlying event.

2. Force as a Description of Response

Classical mechanics gives force a central role. When the motion of an object changes, force provides a quantitative way to connect that change with the object's mass and surrounding interactions. This structure works remarkably well. It allows scientists and engineers to calculate motion, balance, collision, rotation, stress, and stability.

Without force analysis, much of modern construction, transport, machinery, and aerospace engineering would be impossible.

But what does the force equation actually tell us? It establishes a relationship. Given a certain mass and a measured acceleration, an effective net force can be identified. Given known forces and mass, the acceleration can be predicted. The equation organizes the relationship between observable quantities. It does not, by itself, tell us what force is made of.

This distinction is easy to overlook because force often appears on the causal side of the equation. We say force causes acceleration. But force is also inferred from acceleration or interaction. In practice, we frequently know force through what the system does. A spring scale measures force through deformation. A load cell measures it through material response. An accelerometer measures changes in motion. A pressure sensor measures distributed interaction across an area. The instrument does not capture a separate substance called force. It registers a physical consequence calibrated into force units.

This does not make force circular or meaningless. It shows that force is relational. The same relationship can also be represented through other physical quantities. In many mechanical problems, one may work through momentum, impulse, potential energy, or action rather than treating force as the only fundamental language. The choice depends on the structure of the problem.

For a brief collision, momentum change may be more revealing than an instantaneous force. For motion in a conservative system, energy may simplify the analysis. For advanced mechanics, formulations based on energy and action can describe motion without beginning from forces in the Newtonian sense.

This plurality is scientifically revealing. If force were the only direct substance governing motion, why could entire physical problems be formulated successfully through other structures? The answer is that these are different descriptive languages for related physical dynamics.

Newtonian mechanics emphasizes force and acceleration. Lagrangian mechanics emphasizes the difference between kinetic and potential energy. Hamiltonian mechanics organizes evolution through generalized position and momentum. Relativity emphasizes spacetime structure. Quantum theory relies on states, amplitudes, operators, and fields.

The physical consequences agree where the theories overlap, but the conceptual starting points differ. This suggests that force may be one coordinate system within scientific thought rather than the unique foundation of physical reality. It is especially effective at the scale of ordinary bodies and engineering problems. It corresponds well with human experience because we naturally think in terms of pushing and pulling.

But deeper physics often asks a different question. Instead of asking, *What force acts on this object?* it asks:

- What interaction changes the momentum?
- What potential structure shapes the evolution?
- What field exists locally?
- What symmetry constrains the process?
- What geometry determines the natural path?
- What boundary conditions organize the response?

In such formulations, force becomes derived. It appears after the deeper structure has been specified. A charged particle responds to an electromagnetic field. An elastic body responds to deformation of its internal structure. A fluid responds to pressure and density gradients. A freely falling body follows spacetime geometry. Force remains a valid output of the description. But it no longer needs to occupy the deepest layer.

3. Energy Gradients and the Direction of Change

Many physical changes occur because a system is not uniform. One region differs from another. A quantity varies across space. A configuration can evolve toward another configuration. There is a gradient.

A temperature difference drives heat transfer. A pressure difference drives fluid flow. An electrical potential difference drives charge redistribution. A concentration difference drives diffusion. A difference in gravitational potential shapes motion. A stretched spring possesses a configuration from which it can release energy. Across these cases, the system responds to structured difference.

Force can often be related mathematically to the spatial change of potential energy. In simple language, the steeper the potential changes with position, the stronger the corresponding force-like response. This gives force a new interpretation. It is not an independent actor inserted into the system. It expresses the local tendency of the system to change under a non-uniform energetic configuration.

Consider a ball on a slope. Ordinary language says gravity pulls the ball downward. An energy-based description says the ball occupies a configuration with higher gravitational potential relative to lower positions. When constraints allow, the system evolves and kinetic energy increases. Neither description is useless. The force description reveals acceleration along the slope. The energy description reveals the relation between configurations. But the energy picture directs attention away from an invisible pulling agent and toward the structured difference within the system.

The same applies to a spring. When compressed or stretched, the spring's internal material configuration changes. Energy is stored in the deformation. When released, the spring returns toward a lower-energy configuration and produces motion. We may say the spring exerts a restoring force. That is correct. But the restoring force is not a separate entity living inside the spring. It is the measurable expression of how the spring's structured material state changes with deformation.

Likewise, fluid does not flow because "pressure force" travels through it as an object. A pressure gradient means the mechanical conditions vary across the fluid. Local interactions produce net motion. In each case, force summarizes directional response.

Yet gradients themselves should not be reified carelessly. A gradient is a mathematical representation of difference. It does not act as a conscious cause. The physical system has an uneven distribution or changing configuration, and the gradient expresses that structure quantitatively.

This matters because replacing force with gradient does not automatically solve the deeper ontology. We can move from force to potential. Then ask what produces the potential. We can move from potential to field. Then ask what produces the field. We can move from field to geometry or quantum structure. Then ask what supports those. Each conceptual move can deepen explanation, but no new term should be treated as the final actor merely because it sounds more fundamental.

Still, the movement from force to gradients is important. It changes the central image of causation. Instead of objects being pushed by external agents, systems evolve through differences within structured physical conditions. Matter responds because its environment and internal configuration are not uniform.

This is close to the central UED intuition. In a Structured Vacuum Energy framework, matter would be understood as an organized form of the universal substrate. Variations, alignments, and energy gradients within that substrate would shape the responses observed as motion, attraction, repulsion, stress, and field effects. Force would then be a manifestation quantity. It would describe the local measurable consequence of deeper energy dynamics.

This remains a conceptual proposal unless formalized and tested. But it follows a direction already visible in established physics. The deeper explanation is moving away from isolated pushes and toward structured differences.

4. Gravity Without Gravitational Force

Gravity provides the strongest example of force losing its fundamental status. In Newtonian mechanics, gravity is a force of attraction between masses. The description is highly successful. It predicts falling bodies, planetary orbits, tides, and the motion of many astronomical systems with remarkable accuracy. For most ordinary applications, it remains sufficient.

Yet general relativity offers a radically different picture. In Einstein's theory, gravity is not treated simply as a force pulling bodies through fixed space. Matter and energy are associated with the curvature of spacetime, and freely moving bodies follow natural paths within that geometry. A planet does not need an invisible cord attaching it to the Sun. Its path reflects the spacetime structure of the solar system.

A falling object is not necessarily being forced away from natural motion. Free fall is natural motion. The object standing on the ground is the constrained one. This reversal is profound. In ordinary experience, standing still appears natural and falling appears forced. In relativity, free fall is locally unforced, while standing on the ground requires continuous support. The floor prevents the body from following its geodesic path. This is why a person feels weight while standing but feels weightless in free fall.

Einstein's thought experiments involving lifts and accelerating frames helped clarify this equivalence. A person in a sealed accelerating cabin may experience effects similar to gravity. A person in a freely falling cabin may temporarily experience the absence of weight. The difference between gravity and acceleration becomes deeply connected to the structure of reference frames and spacetime.

This was not merely philosophical elegance. General relativity produced testable consequences. It accounted for the unexplained part of Mercury's orbital precession, predicted the bending of light near massive bodies, and described gravitational redshift and other effects. The broader theory succeeded where the force-only picture reached limits.

Yet gravitational force did not disappear from daily physics. Engineers still calculate weight. Astronomers still use Newtonian approximations. Students still solve gravitational force problems. The old description survives because it is efficient within its domain. This shows that a concept can be non-fundamental without becoming false. Gravity-as-force is an effective description. Gravity-as-geometry is a deeper description.

But even geometry may not be the final layer. General relativity and quantum theory remain difficult to unify completely. At very small scales or extreme conditions, spacetime itself may require a deeper explanation. Some approaches treat geometry as emergent. Others seek quantum descriptions of spacetime. No final consensus has closed the question.

This is where scientific caution becomes essential. It would be premature to say that gravity has been explained absolutely because it has been geometrized. Geometry may describe the structure with extraordinary power while still leaving open why spacetime exists, why it curves, and whether it arises from something deeper.

The historical lesson remains clear: what one theory calls force, another theory may call geometry. The observed motion stays. The explanatory category changes. This alone is enough to challenge the claim that force must be fundamental. If gravitational behaviour can be described more deeply without gravitational force as a primary agent, then force may belong to the level of approximation rather than ontology. Gravity teaches us that causation in physics may not always be an object acting on another object. Sometimes what appears to be action is the consequence of structure.

5. Contact Forces Are Field Responses Too

One might accept that gravity is special while still believing that contact forces are obviously fundamental. After all, when a hand pushes a table, the contact is visible. When a hammer strikes a nail, the impact is direct. When a rope pulls a load, the connection is material. Surely these are real forces.

They are certainly real interactions. But even contact becomes field-like at deeper scales. Two objects that appear to touch do not simply merge their matter. Their atoms approach one another, and electromagnetic interactions resist further overlap. Electron structures, quantum restrictions, and the organization of matter produce the solidity we experience.

The table supports the hand because its internal structure responds. The hand deforms. The table deforms. Electromagnetic interactions redistribute through both materials. The macroscopic result is described as a normal force. But the normal force is not an additional substance inserted between the surfaces. It is the large-scale manifestation of microscopic field interactions and material constraints.

Friction follows a similar pattern. At the textbook level, friction is represented as a force opposing relative motion. At the microscopic level, surface roughness, adhesion, deformation, molecular interaction, and energy dissipation all contribute. The familiar friction force compresses these complex processes into a practical quantity. Again, the quantity is real as a measurable relation. But it is not necessarily fundamental.

Tension in a rope is also distributed through material structure. Pull one end, and changes propagate through the rope at a finite speed determined by its elastic properties. The whole rope does not instantly know that it has been pulled. Internal electromagnetic interactions transmit the disturbance. What appears macroscopically as one uniform tension emerges from local responses throughout the material.

A spring force arises from changes in the spring's internal configuration. Elasticity is connected to how atoms and molecular structures shift and resist displacement. Even the sensation of hardness reflects the energetic cost of forcing matter away from stable arrangements.

Thus contact forces do not escape the broader pattern. They too can be explained through fields, structures, energy changes, and local interactions. This does not mean the force concept should be removed from mechanics. Trying to calculate a bridge directly from quantum electrodynamics would be absurdly inefficient. Engineering requires effective quantities suited to the scale of the problem. Force is one of those quantities.

The deeper theory explains why the effective quantity works. The higher-level description allows practical calculation. This layered relationship appears throughout science. Chemistry remains valid even though molecules arise from quantum physics. Biology remains valid even though organisms are made of molecules. Thermodynamics remains valid even though temperature and pressure emerge statistically. Mechanics remains valid even though contact forces arise from microscopic field structure.

The existence of deeper explanation does not destroy the higher level. It clarifies its status. Force is a macroscopic vocabulary for the organized response of matter. At ordinary scales, it behaves with enough stability to be treated as though it were a direct agent. At deeper scales, the agent dissolves into interaction.

This is one reason physics keeps moving from nouns toward relations. The table does not contain a normal-force substance. The rope does not contain tension as a material fluid. The spring does not store a hidden pushing object. Their structures respond to changed conditions. Force is the measurable face of that response.

6. The Changing Meaning of Causation

Human beings naturally imagine causation as a sequence of visible actions. One thing acts. Another thing changes. A hand pushes a ball. The ball moves. The structure is simple:

Agent → Action → Consequence

This form of causation works well in daily life. It is closely connected to intention. People act for reasons. Tools transmit effort. Events appear to have identifiable beginnings. When this model enters physics, force easily becomes the agent. Force acts on matter. Matter accelerates.

But deeper theories complicate this picture. In a field description, the responding object interacts with the local field condition. The distant source may have contributed to creating that field, but the immediate relation is local. In relativity, matter follows geometry. In quantum theory, outcomes are organized through states, amplitudes, interactions, and probabilities. In thermodynamics, macroscopic direction emerges statistically from vast numbers of microscopic configurations.

Causation becomes distributed. It may not belong to a single object or moment. Consider Earth orbiting the Sun. What causes the orbit? A Newtonian answer says gravitational force. A relativistic answer says curved spacetime. An energy-based answer may refer to the system's configuration and conserved quantities. Each answer identifies a different explanatory layer. Which one is the cause? The question may be too simple.

The orbit arises from the total structure of the system: mass-energy distribution, spacetime geometry, initial motion, and continuing dynamical relation. No single noun contains the whole process. Modern physics often replaces linear causation with structural causation. An event occurs because the system is configured in a certain way. The possible evolution is constrained by geometry, symmetry, conservation laws, boundary conditions, and interactions.

This is not causation disappearing. It is causation becoming less like an external command and more like organized consequence. A river flows not because "flow force" commands it, but because elevation, pressure, gravity, terrain, viscosity, and available paths form a dynamic structure. A crystal forms because local interactions and environmental conditions permit stable organization. A wave propagates because disturbances can be transmitted through the structure of a medium or field. A planet orbits because motion unfolds within a gravitationally structured system.

This broader view changes the question from: *"What thing caused this event?"* to: *"What configuration made this consequence possible or necessary?"*

That shift is central to UED. Universal Energy Dynamics treats existence as continuous consequence rather than a sequence of isolated objects delivering forces to one another. In that view, causation is not imported into reality from outside. It emerges from differences, constraints, alignments, instabilities, and interactions within the universal substrate. A local energy delta produces a response. The response changes the surrounding field. The changed field modifies later possibilities. Cause and consequence form a continuous loop.

This may be more appropriate for physical systems than the simple image of one object pushing another. But caution remains necessary. Replacing force causation with "energy dynamics" does not automatically explain everything. The new framework must specify what energy means, how gradients form, how responses propagate, and what measurable predictions distinguish it from existing theories. Otherwise, the phrase becomes only a larger story.

The value of the shift lies in the direction of inquiry. Force-centred causation asks which agent acted. Structure-centred causation asks how the system was organized. The second question may reveal more of what reality is doing.

7. If Force Is Secondary, What Is Fundamental?

Once force is moved away from the foundation, a new problem appears. What takes its place? The obvious answers are fields, energy, geometry, symmetry, information, or quantum states. Modern physics uses all of these. But none should be accepted casually as the final layer.

A field is deeper than force in many theories, yet we may still ask what the field physically is. Energy is conserved and mathematically central, yet it is not always simple to treat as a substance. Geometry explains gravitational motion, yet spacetime may itself emerge from deeper relations. Symmetry organizes physical laws, but one may ask why those symmetries exist. Information has become increasingly important, but information usually requires a physical state or distinction to be information about something.

Every answer opens another question. This does not mean the search is hopeless. It means fundamentality should be approached carefully. A concept is not fundamental merely because the current theory begins with it. It is fundamental only relative to the framework. Newtonian mechanics begins with mass, force, space, and time. General relativity reorganizes those categories. Quantum field theory begins elsewhere. A future theory may reorganize them again.

Perhaps the most durable foundation is not any one quantity, but the existence of structured relational dynamics. Something differs. Something responds. A pattern forms. A configuration stabilizes. A disturbance propagates. A consequence alters later conditions. From this perspective, force is one measurable expression within a larger process.

Universal Energy Dynamics proposes SVE as the continuous structured basis of that process. Within this view:

- Matter is organized substrate.
- Fields are distributed substrate conditions.
- Gradients are structured differences.
- Motion is evolving response.
- Force is a manifestation variable describing that response.
- Causation is continuous consequence within the shared substrate.

This provides conceptual unity. It connects objects, fields, energy, and interaction without requiring force to act as an independent entity. But conceptual unity is only the beginning of physics. A scientific framework must do more than sound coherent. It must reproduce established results where they are reliable. It must clarify unresolved problems. It must generate formal relations. It must expose itself to possible failure.

A substrate theory that explains everything only after the event explains very little scientifically. The real UED challenge is therefore not to declare force unreal. It is to show how force equations emerge from deeper substrate dynamics and where the new account predicts something distinguishable.

Until that work is complete, force remains an indispensable operational quantity. Its practical status is secure. Its fundamental status is open. This is a productive position. We do not need to choose between naïvely treating force as a substance and carelessly rejecting it as fiction. We can recognize force as real at the level of measurable manifestation.

A bridge experiences loads in an engineering sense. A spacecraft changes trajectory. A material deforms. A charged particle accelerates. These consequences can be calculated through force. But beneath the calculation, the physical process may consist of fields, gradients, geometry, momentum exchange, and structured energy response.

Force is then neither illusion nor foundation. It is the middle layer. It connects deeper dynamics with observable motion. That may be its true place in physics.

Closing Reflection: The Floor Beneath Your Feet

Stand again on the floor. You feel your weight. But now the experience has changed. What seemed like the direct sensation of gravity can be understood as the body's response to support. Earth's surface prevents your natural fall. The floor pushes against your feet. Internal stresses pass through the body. Muscles maintain balance. Biological awareness interprets the entire interaction as heaviness.

Gravity remains present. But what you feel is constraint. This small realization contains the whole argument of the chapter. Force is closest to us through experience, yet the experience does not necessarily reveal the deepest physics.

Objects do not consciously feel force. They deform, accelerate, exchange momentum, redistribute energy, and respond to surrounding structure. Science measures those responses. Classical mechanics expresses them through force. Field theories place the response within local field conditions. Energy formulations connect it with gradients and configuration. General relativity explains gravitational motion through geometry. Quantum theory moves toward deeper interaction structures.

At every stage, force remains useful. But its meaning changes. It begins as push and pull. It becomes a measurable relation. It becomes the effect of a field. It becomes a gradient of potential. In gravity, it may disappear into geometry. In contact, it dissolves into microscopic electromagnetic response. The word survives because the macroscopic pattern survives. Yet survival of the word does not prove fundamentality.

Perhaps force is best understood as the language of changed motion and constrained structure. It tells us what a system does under particular conditions. It may not tell us what reality ultimately contains. This does not reduce physics to semantics. The calculations remain real. The consequences remain measurable. A wrong force calculation can destroy a machine or collapse a structure. But the seriousness of the result should not tempt us to turn the quantity into an invisible substance.

Force is effective because it captures stable behaviour. Its effectiveness is practical truth. Its ontology remains open. The deeper physical question is no longer simply: *What force acted on the object?* It is: *What structured condition produced this response?*

Was there a field gradient? A change in potential? A geometric constraint? A momentum exchange? An internal deformation? A deeper substrate variation? These questions move physics from isolated action toward continuous dynamics. They replace the image of nature as a collection of objects pushing one another with a picture of matter responding within structured reality.

The floor remains beneath your feet. You remain still. Yet at the deeper level, nothing is merely still. Matter is structured. Interactions are active. The body is constrained. Earth is moving. Spacetime is shaped. Energy is distributed. The quantity called force gathers one part of this vast process into a form we can calculate.

That may be exactly why force is indispensable. And exactly why it may not be fundamental.

Chapter 6 — Mathematics Is Powerful, but It Is Not Reality

The Bridge and the Calculation

Before a bridge is built, it exists as lines, dimensions, equations, loads, stresses, tolerances, and material specifications. Engineers calculate how much weight it should carry. They estimate wind pressure, vibration, fatigue, thermal expansion, and the behaviour of foundations. Every major component is translated into quantities. The bridge is tested mathematically before it carries a single vehicle.

If the calculations are wrong, the bridge may fail. This gives mathematics an authority that is difficult to challenge. A formula written on paper can determine whether a structure standing hundreds of metres above the ground will remain stable. An equation can guide a spacecraft toward another planet. Mathematical models can predict eclipses, electrical behaviour, fluid motion, and the response of materials under extreme conditions.

Mathematics does not merely describe what has already happened. It allows us to anticipate what will happen. Yet the bridge is not made of equations. Vehicles do not travel across symbols. Concrete does not know the stress formula. Steel does not read the design drawing. The river below does not recognize the dimensions marked on the plan.

The mathematical bridge and the physical bridge are connected, but they are not identical. One is a representation. The other is an organized material reality exposed to changing conditions. The design may be exceptionally accurate, but it necessarily omits something. It simplifies the distribution of material. It assumes ranges of temperature, load, and behaviour. It assigns properties to steel and concrete according to tested standards. It reduces a vast physical system into variables that can be managed.

The bridge succeeds when the representation captures enough of the relevant physical structure. This is the power of mathematics. It can ignore almost everything and still preserve what matters. But that same success can create confusion. Because the equation predicts so well, we may begin to believe that reality is the equation. Because mathematics reveals invisible regularities, we may assume that mathematical existence and physical existence are the same. Because a model is internally consistent, we may treat every object appearing in it as something that must exist in nature.

The bridge warns us against this mistake. The calculation is indispensable. The representation can be more precise than ordinary perception. Yet no equation contains the complete bridge. It does not contain the sound of traffic, the small imperfections in every component, the ageing of material, the choices of workers, the changing river, or the full history of the structure. It selects what can be formalized for a particular purpose.

Mathematics is one of humanity's most powerful ways of seeing. But seeing reality through a formal structure is not the same as possessing reality itself.

1. Why Mathematics Works So Well

The effectiveness of mathematics in physics is astonishing. Numbers and geometrical relations developed through human thought repeatedly correspond to the behaviour of the physical world. Equations created in one context often reappear in another. Abstract structures developed without immediate practical purpose later become central to physical theories.

This remarkable correspondence has often been described as the *"unreasonable effectiveness"* of mathematics. The phrase captures a genuine mystery. Why should symbols written by human beings describe planets, waves, atoms, electricity, and spacetime so accurately? Why should the same mathematical relation apply to phenomena occurring at vastly different scales? Why can a theory predict something not yet observed, only for experiment to confirm it later?

There are several possible answers. One is that reality itself possesses stable structure. Objects and systems do not behave arbitrarily. Similar conditions generate sufficiently similar consequences. Symmetries persist. Ratios recur. Changes follow patterns. Mathematics succeeds because the universe is organized enough to support abstraction.

A second answer is that human beings developed mathematics through interaction with that universe. Early counting arose from objects and groups. Geometry developed through space, measurement, construction, and astronomy. Timekeeping emerged from repeated change. Mathematical thinking did not appear in complete isolation from reality. Our cognitive structures also evolved within the physical world. The intelligence creating mathematics is itself a product of the same reality mathematics attempts to describe. The fit may therefore be less surprising than it first appears.

A third answer is selection. Physics retains mathematical models that work and abandons those that do not. Many equations can be written. Only a small number survive experimental testing. The history of science is full of unsuccessful models, forgotten formulations, and elegant ideas that did not correspond sufficiently to observation. We remember the mathematical triumphs. We rarely teach all the failed attempts.

The effectiveness of mathematics may therefore arise from a combination of three factors:

- Reality contains regular structure.
- Human intelligence is capable of abstracting that structure.

- Scientific testing selects the abstractions that correspond most reliably.

Even so, something extraordinary remains. Mathematics often goes beyond direct experience. It allows physics to describe curved spacetime, quantum amplitudes, field behaviour, and structures that cannot be visualized through ordinary intuition. The equation becomes a new sense organ. It reveals relationships the eyes cannot see and the body cannot feel.

But revelation must not be confused with creation. The fact that mathematics exposes a pattern does not mean the symbols produce the pattern. Planetary motion did not begin when orbital equations were written. Electromagnetic waves did not wait for Maxwell's formalism. Curved spacetime did not appear after Einstein described it. The mathematics identifies structure already active within reality.

This gives mathematics its legitimate authority. It is not merely decorative language. It constrains thought, exposes contradiction, and produces predictions that ordinary words cannot achieve. But its authority is relational. Mathematics works because it corresponds to physical behaviour under defined conditions. Without that correspondence, an equation remains only a formal possibility. A perfectly consistent mathematical world may exist on paper while failing to describe the physical universe.

This distinction becomes crucial when physics moves into regions where direct testing is difficult. Mathematical elegance can become psychologically persuasive. A theory may appear "too beautiful to be false." But beauty is not measurement. Consistency is not confirmation. Possibility is not existence.

Mathematics is powerful because it reveals order. Science remains science because reality is allowed to answer back.

2. Equations Are Compressed Descriptions

An equation is one of the most efficient forms of human knowledge. A short expression can condense years of observation, experiment, argument, and conceptual development. It can connect quantities across situations and allow consequences to be calculated without repeating the entire history of discovery.

But an equation is compressed. Compression always leaves something out. Consider a simple relation describing the motion of an object under constant acceleration. The equation may connect initial velocity, acceleration, time, and final velocity. It is precise within its assumptions. But it does not tell us why the acceleration is constant. It does not identify the mechanism producing it. It does not describe the material composition of the object. It does not include air resistance unless that is added. It does not reveal whether the object is a stone, vehicle, droplet, or spacecraft.

The equation selects a limited pattern. That is exactly why it is useful. If the equation attempted to include every property of the event, it would become unmanageable. Scientific formalism gains power by separating relevant variables from irrelevant detail.

The same applies to more advanced equations. Maxwell's equations capture electromagnetic relations without describing every material arrangement in which electromagnetic effects occur. Einstein's field equations connect matter-energy with spacetime geometry, but their application requires specific conditions, assumptions, and boundary values. The Schrödinger equation organizes quantum evolution, but it does not, by itself, settle every question about measurement or the ontology of the wavefunction.

The equation provides structure. Interpretation determines what the structure means physically. This is why the same equation can sometimes support different conceptual pictures. Mathematics specifies relations among symbols. Physics assigns those symbols to measurable quantities or theoretical entities. That assignment is not automatic. The symbol x does not know whether it represents position, concentration, displacement, or something else. The equation becomes physical only when its terms are connected with observations and operational definitions.

This creates several layers:

- **Formal structure** — the mathematical relation.
- **Physical interpretation** — what the terms are said to represent.
- **Operational definition** — how those terms are measured.
- **Domain of validity** — the conditions under which the relation works.
- **Experimental consequence** — what the equation predicts.

When these layers agree, the equation becomes a powerful physical law or model. When they are confused, mathematical symbolism may hide conceptual uncertainty. A formula can appear complete because every symbol is defined. Yet the definitions may still rest on unresolved assumptions.

For example, a field may be represented mathematically at every point in space. The equation can describe how the field changes. But the formalism alone does not settle whether the field is a substance, a relation, a state of spacetime, or an emergent manifestation. The equation tells us how the representation behaves. It may not tell us what the represented reality ultimately is.

This does not reduce equations to mere shorthand. Compression can preserve deep structure. A musical score is not the sound of an orchestra, yet it preserves enough organization for the music to be reproduced. A genetic sequence is not the living organism, yet it carries essential instructions. A map is not the landscape, yet it can guide movement.

An equation similarly preserves relational architecture. It can reveal what remains constant while physical situations change. It can unite many events under one pattern. But the equation remains a representation. Its greatness lies not in being reality, but in capturing reality with extraordinary economy.

3. When Mathematics Predicted What No One Had Seen

The strongest argument for the physical power of mathematics comes from successful prediction. Sometimes an equation is developed to organize known observations and then produces consequences no one originally placed into it. Those consequences are later confirmed. This is difficult to dismiss as mere curve fitting.

Maxwell's mathematical unification of electricity and magnetism implied the existence of electromagnetic waves. The calculated propagation speed matched the known speed of light, suggesting that light itself was electromagnetic. The mathematics revealed a connection deeper than the separate observations from which it began.

General relativity predicted effects that were not part of ordinary Newtonian thinking, including the bending of light by gravity and particular changes in time and frequency within gravitational conditions. Quantum theories predicted particles and phenomena before they were experimentally established. Dirac's equation, developed to describe electrons consistently with relativity and quantum principles, led toward the prediction of antimatter. What first appeared as an implication of mathematical structure later gained experimental support.

Such episodes give mathematics a nearly prophetic quality. The equation seems to know more than the scientist who wrote it. But we should examine what this really means. The mathematics does not think. It contains consequences implicit within its assumptions and structure. Human reasoning may not immediately notice all of them.

Once the formal framework is constructed, deduction exposes relationships that were hidden from intuition.

The prediction succeeds because the framework has captured something real about the physical system. This is one of the highest achievements of theoretical physics. Yet even successful prediction does not prove that every element of the mathematical interpretation corresponds literally to an independent object in nature.

A theory may predict correctly because its relational structure is accurate, even if the picture attached to it later changes. Newtonian gravity predicted planetary motion with extraordinary success. Its force interpretation was later placed within a broader geometric account. Maxwell's theory predicted electromagnetic waves, while the mechanical ether often associated with wave propagation did not survive in its classical form. Quantum equations predict experimental outcomes extremely well, while physicists continue to debate what the formalism implies about reality.

The predictive success belongs most securely to the relational structure. The ontology may remain provisional. This distinction becomes especially important when mathematics predicts entities that cannot yet be observed directly. A mathematical solution may represent:

- a physically real object;
- an unstable state;
- an unphysical branch excluded by conditions;
- an artefact of approximation;
- a possibility that nature does not realize;
- or a clue that the theory requires reinterpretation.

Not every formal solution is a physical discovery. Physics must ask whether the solution corresponds to measurable consequence. This is why experiments remain essential. Mathematics can move ahead of observation, but science requires the two to meet eventually. A theory that never risks contact with evidence may remain mathematically sophisticated while drifting into speculative metaphysics.

The great predictive episodes in physics should therefore inspire both confidence and restraint. Confidence, because formal reasoning can uncover structures far beyond ordinary intuition. Restraint, because predictive success confirms specific consequences, not automatically every philosophical story surrounding the theory.

Mathematics can reveal what follows from a model. Experiment determines whether nature follows that path. And interpretation asks what the agreement means. These are related activities. They are not the same.

4. Mathematical Beauty and the Temptation of Reality

Physicists often value elegance. A compact theory that unifies many phenomena is usually preferred to a complicated system of unrelated corrections. Symmetry, simplicity, and mathematical coherence have repeatedly guided major advances. This preference is not superficial. Nature often does display deep regularity. Conservation laws connect with symmetries. Different forces can be unified under larger structures. Complex behaviour may emerge from simple principles.

Mathematical beauty can therefore function as a valuable research guide. But a guide is not evidence. A theory may be beautiful and wrong. A model may be ugly and accurate. A simple equation may work only within narrow conditions. An elegant unification may rely on assumptions that nature does not realize.

The difficulty is psychological. When a mathematical structure is exceptionally coherent, it creates a feeling of inevitability. Every part fits. Contradictions disappear. Different phenomena align under one formal scheme. The theory begins to feel discovered rather than constructed. Researchers may come to believe that reality must possess the same elegance. This belief has sometimes been productive. It has also sometimes prolonged commitment to ideas that lacked experimental support.

Institutions can amplify the effect. Once a mathematically rich framework attracts researchers, funding, conferences, and academic prestige, it develops its own intellectual ecosystem. Problems are defined within it. Careers are built around it. New mathematical refinements appear. The framework becomes increasingly sophisticated even if decisive empirical confirmation remains absent.

This does not make the research invalid. Exploration of mathematically possible structures can be valuable. It may generate tools useful elsewhere or reveal hidden connections. The problem arises only when mathematical development is presented as equivalent to physical confirmation.

Physics must preserve a boundary between three statements:

- The mathematics is consistent.
- The mathematics describes existing observations.
- The entities implied by one interpretation physically exist.

These are not interchangeable claims. A mathematically consistent theory may fail observationally. A theory may fit observations while remaining one of several possible descriptions. A theoretical entity may be useful within calculation without possessing simple object-like existence.

The language used in popular physics often collapses these distinctions. Readers hear that extra dimensions exist, the universe split, particles borrowed energy, or spacetime emerged from information. Sometimes these phrases represent serious theoretical proposals. But the grammar often removes the word *may*. Mathematical possibility becomes physical fact in one sentence.

This is how sophisticated physics can become mythology without intending to. The story is presented with equations behind it, so it appears experimentally secured. But equations secure only what has been tested through their consequences. The rest remains interpretation.

This caution must also apply to UED. Structured Vacuum Energy should not be treated as physically established merely because it can organize concepts coherently. A universal substrate may offer an attractive picture. It may connect fields, matter, gradients, and manifestation. But conceptual elegance is not enough. UED must produce mathematics that corresponds with known evidence and leads toward distinctive tests. Otherwise, SVE risks becoming exactly what this book criticizes: a compelling ontological story promoted beyond its evidential status.

A serious new framework must therefore welcome mathematical discipline without surrendering to mathematical seduction. The equation should clarify the proposal. It should expose assumptions. It should generate consequences. It should make failure possible. Beauty may invite investigation. Only reality can grant acceptance.

5. What Mathematics Reveals

Mathematics reveals relationships that ordinary language cannot hold with sufficient precision. It reveals proportion. It identifies rates of change. It exposes invariance. It tracks how one variable depends on another. It distinguishes linear from nonlinear behaviour. It describes probability distributions, oscillations, curvature, symmetry, feedback, and stability. It can show that two apparently different phenomena share the same underlying form.

This is one of its greatest contributions to physics. The eye sees a swinging pendulum, a vibrating spring, a sound wave, and an electrical oscillation as different events. Mathematics may reveal a shared pattern. The physical materials differ. The relational structure repeats.

This capacity allows physics to travel across domains. A mathematical method developed for one problem may later describe another. Differential equations used for mechanical motion may appear in electrical circuits, waves, diffusion, population dynamics, and quantum theory. Mathematics reveals the geometry of change. It does not depend on the surface appearance of the system. It asks what relation persists beneath the visible differences.

Mathematics also reveals hidden consequences. A theory may imply conservation laws that are not obvious from individual observations. A symmetry may show why certain changes are impossible. A small instability in an equation may predict large later consequences. A model can identify threshold behaviour that ordinary intuition would miss.

In this sense, mathematics expands human intelligence. We cannot mentally follow every molecule in a gas, but statistical mathematics allows collective behaviour to be understood. We cannot visualize four-dimensional spacetime directly, but geometry allows its relations to be calculated. We cannot observe a quantum state in the way we see a stone, but mathematical formalism connects preparation with measurable probabilities.

Mathematics also forces consistency. Ordinary language can hide contradiction through ambiguity. Equations are less forgiving. If assumptions conflict, the formal structure may break. If units do not match, the relation is suspect. If a prediction diverges from measurement, revision becomes necessary. This discipline is invaluable. Philosophy without mathematical constraint can become vague. Mathematics without physical interpretation can become detached. Physics gains strength from their cooperation.

Yet what mathematics reveals is always structural. It shows how quantities and formal objects relate within a defined framework. It does not automatically reveal meaning. An equation can show that a quantity is conserved. It cannot, by itself, explain why conservation is fundamental. A symmetry can generate a mathematical constraint. It does not necessarily tell us why reality possesses that symmetry. A wavefunction can evolve precisely. The equation alone may not tell us what the wavefunction is.

Mathematics can reveal that a pattern exists. It may not reveal what exists as the pattern. This is a subtle but critical distinction. Suppose a mathematical model predicts a field value at every point. The formalism reveals distribution and change. But the ontological question remains: *What physical reality does the field value represent?* A real medium? A probability? A geometric relation? A computational device? A state of a deeper substrate?

The answer requires interpretation, experiment, and perhaps a broader theory. Mathematics reveals the architecture of a description. When the description corresponds with nature, it reveals architecture within physical behaviour. But whether that architecture is fundamental, emergent, approximate, or representational remains a further question.

Mathematics opens the door. It does not always tell us what room we have entered.

6. What Mathematics Cannot Decide Alone

Mathematics cannot determine, by itself, which physical world exists. Many internally consistent mathematical systems are possible. Nature appears to realize only some of them, or perhaps one particular structure among many conceivable alternatives. Formal consistency is therefore necessary for a physical theory, but not sufficient.

Mathematics cannot decide alone which quantities should be measured. That choice arises from observation, experimental possibility, and conceptual judgement. It cannot decide alone which idealization is appropriate. A model may treat matter as continuous, discrete, rigid, elastic, classical, relativistic, or quantum depending on the problem. The physical context determines which representation is useful.

Mathematics cannot decide the interpretation of every symbol. The same formal structure may support different physical readings. This is especially evident in quantum mechanics, where multiple interpretations can reproduce the same experimental predictions. The mathematics may be common. The ontologies differ.

Mathematics also cannot tell us whether a theoretical entity exists merely because it appears in an equation. Coordinates appear in equations, but a coordinate label is not a physical substance. Gauge choices may alter mathematical representation without changing observable physics. Virtual entities may be useful in calculations without behaving like directly detectable particles. Some infinities indicate physical limits; others expose a breakdown in the model. Formal objects must be interpreted according to their role.

Mathematics cannot supply meaning independently of consciousness. The symbol acquires meaning through definition and use. An equation does not know what it represents. It does not know whether it describes a bridge, star, market, population, or abstract space. The same mathematical relation can model many systems because mathematics captures form rather than lived identity. This is both its strength and its limitation.

Mathematics also cannot replace experiment. A calculation may show what would happen if the assumptions are correct. Only observation can test whether the assumptions correspond to physical conditions. Even highly successful mathematics can fail when extended beyond its proper domain. Newtonian mechanics works extraordinarily well but not universally. Euclidean geometry is powerful but not sufficient for curved spacetime. Classical probability cannot simply replace quantum formalism. Continuum mechanics eventually reaches molecular limits. Every mathematical model has conditions under which its representation is adequate.

Perhaps most importantly, mathematics cannot answer every question beginning with *why*. It may show why one result follows from another within the theory. But when asked why the fundamental structure itself exists, mathematics often reaches a boundary. Why these laws? Why these constants? Why this dimensional structure? Why is reality mathematically describable? A deeper theory may reduce some assumptions, but it introduces others. At some point, explanation meets foundation.

This does not mean mathematics has failed. No tool should be criticized for not performing a task outside its nature. A microscope does not explain morality. A thermometer does not reveal consciousness. Mathematics reveals relation, structure, consistency, possibility, and consequence. It does not, by itself, convert possibility into existence or representation into meaning.

To do physics responsibly, we need mathematics, observation, experiment, conceptual clarity, and philosophical restraint. Remove mathematics, and physics loses precision. Remove observation, and it loses contact with nature. Remove philosophy, and it may not recognize the assumptions hidden inside its symbols.

7. Truth, Representation, and the Levels Between

The word *truth* can become dangerous when applied too simply to scientific theories. Is Newtonian mechanics true? Yes, within the conditions where it accurately describes motion. Is general relativity true? Its predictions have received extensive support, and it describes gravity more broadly than Newtonian theory. Does this make Newton false? Not in the ordinary domain where it remains an excellent approximation.

Scientific truth is often layered. A theory may be operationally true without being fundamentally complete. It may represent one scale accurately while emerging from a deeper scale. It may preserve the right relations while using a picture later reinterpreted. Temperature is real at the thermodynamic level even though no single molecule possesses the temperature of the entire system. Pressure is real at the macroscopic level even though it emerges from microscopic interactions. A solid object is real at the human scale even though its deeper physical structure does not resemble a perfectly continuous block.

Truth in physics is therefore not always a choice between total reality and complete illusion. There are levels of valid representation. A map can truthfully represent roads while omitting geology. Another map can truthfully represent elevation while distorting roads. Neither contains the complete territory. A physical theory works similarly. It highlights certain structures because those structures answer particular questions.

The danger begins when a level-specific truth is promoted into an absolute ontology. Newtonian force becomes the final cause. Spacetime geometry becomes unquestionably fundamental. Quantum fields become the ultimate substance. Information becomes the foundation of everything. A substrate becomes the final answer. Each promotion may exceed what the evidence establishes.

A more disciplined position is: *This representation captures these manifestations under these conditions.* This phrasing may appear modest, but it is scientifically strong. It states what is known without pretending that the unknown has disappeared.

The distinction between truth and representation also protects science from two opposite errors. The first is dogmatism. Dogmatism assumes that the present theory has identified reality exactly as it is. The second is cynicism. Cynicism assumes that because theories are representations, they are merely arbitrary stories. Neither is justified. Representations can be objectively better or worse. A theory that predicts accurately, unifies observations, survives testing, and reveals new consequences is not equivalent to an unsupported narrative. The territory constrains the map. But even the best map remains a map.

This distinction is essential for the future development of UED. If UED is presented as final truth, it closes the very inquiry it claims to reopen. If it is presented only as metaphor, it cannot become physics.

Its proper path lies between these extremes. UED should offer a structured representation of reality in which SVE, energy dynamics, matter formation, fields, and manifestation are connected. The representation should recover established results where they work and identify areas where new predictions arise.

Its truth will not come from philosophical confidence. It will come from correspondence, coherence, and consequence. Mathematics will be necessary because it turns broad intuition into accountable structure. But the equations, if developed, will still remain representations. They will not become SVE itself. No symbol becomes the substrate it describes.

This is the deeper discipline required of every physics theory: trust the representation where it works. Do not confuse its success with complete possession of reality.

Closing Reflection: The Equation and the World

Return to the bridge. Before construction, its mathematical representation may contain every dimension required by the engineer. Loads are calculated. Materials are specified. Safety factors are included. Simulations test wind, traffic, vibration, and fatigue. The design may be so precise that the finished bridge behaves almost exactly as predicted.

Yet when construction begins, the equation enters a world it does not control. Materials vary. Weather changes. Workers make decisions. Foundations meet actual geology. Components age. Traffic exceeds expectations. The representation must continually answer to the physical bridge.

This is the relationship between mathematics and reality. Mathematics does not become less powerful because it is a representation. Its power comes precisely from representation. It isolates structure. It compresses complexity. It reveals relation. It allows prediction before action. It turns invisible patterns into forms intelligence can manipulate.

But the bridge is still more than the calculation. The universe is still more than the equation. Nature does not contain symbols written in human notation. It contains the stable relationships that make those symbols effective. An equation may reveal those relationships with astonishing accuracy. It may predict something no one has seen. It may unify phenomena once thought unrelated. It may expose an error in intuition. It may guide technology across enormous distances and microscopic scales.

Yet it cannot interpret itself. It cannot decide whether every mathematical entity physically exists. It cannot determine which possible world is the actual world without evidence. It cannot turn elegance into confirmation. It cannot replace the territory with the map.

Mathematics reveals the structure of what can be represented. Physics asks whether that representation corresponds to what happens. Philosophy asks what the correspondence means. Experiment allows reality to respond. All four are necessary when inquiry reaches the foundations.

The proper lesson is therefore not that mathematics is limited and should be distrusted. The lesson is that its limits should be understood so its real power can be respected. An equation is not reality. But it can reveal a stable doorway into reality. A theory is not the universe. But it can organize genuine features of the universe. A mathematical object is not automatically a physical object. But it may point toward something no instrument has yet detected.

The distinction between truth and representation is not a weakness in science. It is what keeps science open. It allows Newton to remain useful after Einstein. It allows thermodynamics to remain valid after molecular theory. It allows fields to be powerful without declaring them final. It allows UED to propose a substrate without demanding belief before evidence.

Mathematics is powerful because reality has structure. But reality is not required to fit every structure mathematics can imagine. The final authority remains the physical world. The equation speaks. Experiment listens. Nature answers.

Interlude II

After the Equation

The problem has been solved. The values were substituted correctly. The units were checked. The graph matched the expected curve. The prediction agreed with observation. The equation worked.

What remains?

In most classrooms, nothing remains. The answer has been obtained, so the problem is considered complete. But in physics, the completion of a calculation does not always complete the event.

The equation may tell us how position changes with time. It may predict the energy of a system. It may relate pressure, volume, and temperature. It may describe the curvature of spacetime. It may assign probabilities to possible quantum outcomes. It may reproduce behaviour with remarkable accuracy.

Still, another question waits behind the result: *What has the equation actually captured?*

Mathematics and Precision

Mathematics gives physics its precision. Without it, scientific language would remain too flexible, too personal, and too dependent on intuition. Mathematics allows one scientist to state a relationship in a form another scientist can test. It reveals consequences that ordinary reasoning may miss. It compresses patterns. It separates relevant variables. It exposes symmetry. It allows prediction. It makes engineering possible.

There is no serious future for physics without mathematics. But there is also no guarantee that mathematical success automatically reveals the whole of physical reality.

An equation expresses structure. It relates symbols according to defined rules. Those symbols acquire physical meaning because they are connected with measurement, experiment, and interpretation.

The equation itself does not contain the falling stone, the vibrating atom, the expanding gas, the curved spacetime, or the detector event. It contains a disciplined representation of selected relations within those phenomena.

Representation and Reality

This distinction may seem obvious until a mathematical model becomes extremely successful. Then the representation begins to appear more certain than the reality it was designed to describe.

We stop saying that the equation represents the system. We begin speaking as though the system *is* the equation.

We say that nature is mathematical. We say that the universe computes. We say that reality is information. We say that particles are wavefunctions. We say that spacetime is geometry.

These statements may express profound insights. But each also carries an interpretive step.

- The mathematics may show that a geometrical model predicts gravitational behaviour. It does not, by itself, settle whether geometry is the final substance of reality.
- The formalism may assign a quantum state to a system. It does not, by itself, settle what the state physically is.
- The equations may preserve information under certain transformations. They do not automatically prove that information exists independently as the foundation of everything.

The mathematical structure is clear. The ontology remains open.

Maps, Scores, and Drawings

This does not reduce the achievement.

A map can guide a traveller accurately without containing the soil beneath the road. A musical score can preserve structure without producing sound by itself. An architectural drawing can specify relations without becoming the building. A mathematical theory can reveal something real without becoming identical to the reality revealed.

The equation succeeds because reality contains stable relations. That correspondence is extraordinary. But correspondence is not identity.

The Student Who Asks Too Much

This is where the student who asks one question too many becomes important.

The calculation gives the correct acceleration. The student asks what causes the acceleration. The teacher points to force. The student asks what force is. The teacher points to a field. The student asks what the field is. The teacher points to an equation. The student asks what physical reality the equation represents.

At this point, the usual chain of explanation begins to weaken. The teacher may become impatient. The student may be told that such questions are philosophical rather than scientific. But the history of physics suggests otherwise.

The movement from force to field was not created by calculation alone. It required a new picture of what could exist in the region around an object. The movement from absolute space and time to relativity required more than adding terms to old formulas. It required questioning what space and time meant. The movement from classical particles to quantum theory required physics to reconsider position, state, probability, and measurement.

Each transformation began where the existing mathematics remained powerful but its interpretation became insufficient. Physics often changes most deeply after the equation has already succeeded.

After Success

The success creates a stable platform. From that platform, a deeper question becomes visible.

- How can the same relation be understood more fundamentally?
- What assumptions were hidden inside the original formulation?
- Which quantity appeared primary only because the deeper structure had not yet been found?
- Why does the mathematical form work?
- What physical architecture makes that form possible?

These questions move physics from calculation toward manifestation.

Manifestation Science

A quantity does not arise from nowhere. Before it can be measured, something physical must happen.

A body must accelerate. A material must deform. A detector must register a signal. A field must vary. Radiation must arrive. A pattern must appear.

The equation relates quantities after manifestations have been identified and translated into measurable form. But what creates the manifestation?

- What makes a pressure reading possible?
- What physical organization makes a solid resist compression?
- What structure allows an electromagnetic wave to propagate?
- What conditions produce a localized particle-like detection?
- What turns microscopic dynamics into macroscopic stability?

The equation may describe the relationship among the measurable variables. The manifestation question asks how those measurable variables become physically available in the first place.

This is a wider task. It does not reject quantitative physics. It surrounds it.

Examples of Manifestation

Consider **temperature**. An equation may relate temperature to pressure and volume. Another may connect it with average energy distributions. The mathematical relation can be precise. But temperature does not appear in nature as a separate object. It becomes meaningful at a certain level of organization. It manifests through collective behaviour. The quantity is real. The underlying architecture is deeper.

Consider **force**. The equation may relate force to acceleration. Another formulation may connect force with a potential gradient. A field theory may describe the local conditions. General relativity may replace gravitational force with geometry. The measurable effect remains. The explanatory layer changes.

Consider **quantum detection**. The mathematical formalism may predict the probability distribution with extraordinary accuracy. The detector records a localized event. Between the formal state and the observed manifestation lies one of the deepest interpretive questions in physics. The calculation works. What physical process has become visible?

The Boundary After the Equation

This is the boundary after the equation. It is not the boundary of usefulness. It is the boundary of representation.

Mathematics tells us how the measured aspects are related. Manifestation science asks how reality becomes measurable through those aspects. The shift is subtle but important.

Instead of asking only *how much force acts*, we ask *what physical response has been named force*. Instead of asking only *how much energy a system has*, we ask *what state and capacity the quantity represents*. Instead of asking only *which field equation applies*, we ask *what kind of reality a field describes*. Instead of asking only *what probability is predicted*, we ask *how a possible state becomes an actual recorded event*.

The language of quantities remains necessary. But it becomes transparent. We begin to see the physical process through it.

Limits of Specialization

This is also where the limits of specialization become visible. A physicist may become highly skilled in solving a particular class of equations. A computer may process more variables than any person can follow. An artificial intelligence system may identify mathematical patterns hidden in enormous datasets.

These abilities will transform scientific work. But calculation alone cannot decide which question matters.

- It cannot determine whether a variable has been reified.
- It cannot automatically separate a formal solution from a physical possibility.
- It cannot tell us whether one successful interpretation has been mistaken for the only possible ontology.

The future of physics will therefore require more than increased computational power. It will require conceptual judgment.

Someone must ask what the symbols represent. Someone must distinguish model from event. Someone must notice when an effective quantity has been promoted into a fundamental substance. Someone must ask whether different theories are describing separate realities or different manifestations of one deeper structure.

The Balance of Inquiry

This is not a retreat into vague philosophy. It is a demand for greater scientific clarity.

- Philosophy becomes dangerous only when it ignores evidence.
- Mathematics becomes dangerous only when its interpretation is hidden.
- Experiment becomes limited only when the observed signal is mistaken for a complete explanation.

Each mode of inquiry needs the others.

Observation gives contact. Measurement gives comparison. Mathematics gives structure. Theory gives organization. Philosophy examines meaning. Experiment returns the idea to reality.

The process should remain open in both directions. Reality is translated into quantities. Quantities are connected through equations. Equations generate predictions. Predictions return to experiment. But after the result is confirmed, interpretation must return once more to reality.

Beyond the Equation

What has been learned about the world itself? The answer may be less complete than the calculation. That is acceptable. Science does not become weak by admitting that the map is incomplete. It becomes weak only when it confuses confidence in the map with possession of the territory.

The chapters ahead will therefore move beyond a purely quantitative definition of physics. Not away from measurement. Not away from equations. Not away from established theory. Beyond them in scope.

Physics may be the science of measurable relations. But it may also be the study of the structures, dynamics, and consequences through which measurable relations emerge.

This means paying attention to:

- Pattern before quantity
- Interaction before formula
- Organization before property
- Manifestation before measurement

The visible answer may be the final line of the numerical problem. It is not necessarily the final line of physics.

The equation has spoken. Its prediction has survived. Now another form of attention is required.

Look again at the event. What changed? What persisted? What became observable? What deeper relation was compressed into the quantity? What part of the explanation belongs to measurement? What part belongs to mathematics? What part belongs to interpretation? What remains unknown?

After the equation, the universe is still there. The object still falls. The field still varies. The material still responds. The detector still records. The pattern still demands meaning.

The calculation has ended. Reality has not.

And from this point onward, physics must become more than the study of quantities. It must begin to study how reality manifests through them.

Chapter 7 — Physics Is More Than Quantities

Before Anyone Measured the Rainbow

A rainbow appeared long before anyone measured the wavelength of light. It formed when sunlight, water droplets, geometry, and the position of an observer entered a particular relationship. Colours separated across the sky. The arc appeared, changed, and disappeared as conditions changed.

For ancient observers, the rainbow was an event before it was an optical phenomenon. It could be interpreted as a sign, a message, a mystery, or an object of wonder. Later science asked different questions. Why does it form? Why is it curved? Why do the colours appear in a particular order? Why does its position depend on the observer? How can the angles be measured?

Eventually, reflection, refraction, dispersion, wavelength, and electromagnetic theory provided increasingly precise descriptions. The rainbow became calculable. But calculation did not create it. Nor did calculation exhaust it. The measurable quantities explained important aspects of its formation, yet the phenomenon remained larger than any one number. No single wavelength contained the rainbow. No angle alone described the full event. The rainbow arose from a structured relationship among light, droplets, space, motion, and observation.

This is how physics often begins. Something manifests. A pattern appears. A consequence is noticed. Only afterward do quantities enter. The order matters. Modern education can make physics look as though it begins with numbers. Students are introduced to units, formulas, graphs, and numerical problems. Success is often measured by how quickly a value can be calculated.

But nature does not first produce a number. It produces an event. The number is our way of identifying one measurable feature within that event. A stone falls before acceleration is calculated. A wave travels before frequency is assigned. A body warms before temperature is recorded. A material deforms before stress and strain are plotted. A star emits light before its spectrum is analysed. Reality manifests first. Physics responds by developing quantities.

This means physics cannot be reduced to quantitative science alone. Quantification is one of its most powerful methods, but it is not its whole subject. Physics also studies:

- what patterns exist;
- how structures form;
- why systems respond differently under different conditions;
- how one consequence becomes the condition for another;
- how stable forms emerge from changing dynamics;
- how scales alter what can be observed;
- how measurable quantities arise from deeper processes.

These questions are not outside physics. They are often the reason quantities are introduced in the first place. The danger begins when the instrument becomes more important than the phenomenon, when the number replaces the event, and when what cannot yet be measured is treated as though it cannot exist.

A broader physics does not reject numbers. It remembers what numbers are for. They are not reality's substance. They are handles through which intelligence grasps selected manifestations of reality. The rainbow is measurable. But it is not made of measurements.

1. Beyond Numbers

Physics is frequently described as the most quantitative of the sciences. The description is justified. Physics depends heavily on measurement, mathematical relation, precision, repeatability, and prediction. Its quantities allow scientists to compare events occurring in different laboratories, countries, centuries, and regions of the universe. Without numbers, much of physics would remain impressionistic.

A person may say that one object moved quickly and another slowly. Physics asks how quickly. A material may feel hot. Physics asks what thermal state can be measured. A structure may appear strong. Physics asks what load it can withstand. The movement from qualitative impression to quantitative description is one of science's greatest advances.

Yet the phrase *"quantitative science"* can become misleading when it suggests that physics studies only what has already been converted into numbers. Before something can be measured, it must first be conceptually identified. A scientist must recognize that a difference exists. A pattern must become visible. A variable must be separated from the surrounding complexity. A method must be created for comparing one case with another. Only then can a quantity be assigned.

Numbers enter after perception, classification, and conceptual invention. Even a simple measurement therefore contains a prior act of intelligence. Suppose a pendulum swings. One observer notices its repeated motion. Another notices that longer pendulums swing more slowly. A third asks whether the mass of the bob matters. Someone then decides what to measure: length, time, angle, amplitude, or energy. The event contains many possible quantities. Physics selects those relevant to the question.

This selection is not arbitrary, but neither is it dictated automatically by the pendulum. The system does not announce which variable should become central. Human intelligence isolates the structure that seems to govern the pattern. This is why important advances in physics often begin not with better calculation, but with a better question.

Galileo's achievement was not merely measuring falling bodies. It involved changing how motion was conceptualized. Faraday's contribution was not only recording electromagnetic effects. He began thinking of the surrounding region as physically structured. Einstein did not simply insert new values into Newtonian equations. He reconsidered the meaning of simultaneity, acceleration, gravity, space, and time.

In each case, the conceptual shift came before the mature formalism. Physics moved forward because someone saw the phenomenon differently. This suggests that the true scope of physics includes at least three activities:

- **Recognition** — noticing a stable or surprising pattern.
- **Representation** — developing concepts and quantities to describe it.
- **Formalization** — expressing the relationship mathematically.

Quantitative physics is mainly associated with the third activity, but it depends on the first two. A number without a concept is meaningless. A measurement without a question is directionless. An equation without a physical interpretation is only formal structure.

Physics therefore cannot be reduced to calculation, even when calculation is indispensable. A student who can solve equations but cannot identify the physical process has learned technique without understanding. A scientist who can fit data but cannot distinguish correlation from mechanism has produced a model without sufficient explanation. An institution that rewards only quantifiable output may overlook the conceptual insight from which the next theory could emerge.

To move beyond numbers does not mean becoming less scientific. It means recognizing that numbers occupy one layer within a broader process of knowing. Physics begins with reality, not arithmetic.

2. Patterns Exist Before Measurements

A pattern does not begin when it is measured. Day and night alternated before clocks. The Moon changed appearance before calendars. Seasons repeated before temperature records. Tides rose and fell before gravitational theory. Planets followed regular paths before telescopes or orbital equations.

Human beings first recognized recurrence. Measurement came later. This sequence reveals an important foundation of science: pattern recognition precedes quantification.

Early observers of the sky did not possess modern instruments, but they noticed repetition. Certain stars returned to familiar positions. The Sun's path changed across the year. The phases of the Moon followed a cycle. Eclipses, though rare, belonged to an order that could eventually be anticipated. These patterns encouraged counting, comparison, and geometry. Astronomy did not begin because someone already possessed equations. Equations became possible because order was first noticed.

The same applies in experimental physics. A phenomenon may initially appear as an anomaly—a repeated deviation, an unexpected trace, a failure of an accepted prediction. At first, there may be no established quantity that captures it well. Scientists must decide whether the pattern is real, instrumental, accidental, or conceptually important.

This stage can be uncertain. New phenomena often appear at the boundary of existing measurement. Instruments were designed according to current expectations. If reality behaves outside those expectations, the first signal may look like error. Institutions therefore face a genuine dilemma. If every irregular result is treated as a discovery, science becomes flooded with noise. If every irregular result is dismissed, important phenomena may be lost. Scientific judgement is required before stable quantification becomes possible.

Pattern recognition is not a lower or primitive stage of science. It is an active form of reasoning. A pattern implies that events are not fully isolated. Something persists across changing conditions. A relation repeats. A system maintains coherence. A deviation carries information.

Patterns may be spatial, temporal, statistical, geometric, or relational. A crystal displays spatial order. An orbit displays temporal and geometric regularity. A spectrum displays a distribution of frequencies. An interference pattern reveals structured superposition. A phase transition reveals a sudden reorganization of collective behaviour. None of these is merely a number. Each is an arrangement.

Quantities are then introduced to characterize the arrangement. Wavelength measures part of a wave pattern. Period measures recurrence in time. Correlation measures relation between variables. Order parameters describe changes across phases. But the pattern is conceptually prior.

This helps explain why visualizations remain so important in physics. Graphs, diagrams, tracks, images, and distributions often reveal structure that a list of numbers conceals. The numbers are necessary. The pattern gives them meaning.

Modern data science can process enormous datasets, but a meaningful discovery still requires identifying which structure matters. An algorithm may detect correlations, yet interpretation remains necessary to determine whether the pattern reflects causation, bias, noise, or a deeper mechanism.

The same issue exists in theoretical physics. A mathematical pattern may suggest a symmetry or unification. But the physicist must ask whether this structure corresponds to nature or only to the chosen formalism.

Physics therefore begins neither with pure observation nor pure mathematics. It begins when intelligence detects stable consequence within change. The universe manifests order. Measurement converts selected aspects of that order into quantities. The sequence should never be reversed.

3. Manifestation: How Reality Becomes Observable

The word *manifestation* refers to the way an underlying physical condition becomes observable through consequence.

We do not directly observe gravity as an object. We observe falling bodies, orbital motion, tides, changes in clocks, bending of light, and other consequences associated with gravitational structure. We do not directly see an electric field as a coloured substance. We observe charge movement, sparks, current, attraction, repulsion, and measurable responses in instruments. We do not see temperature itself. We observe expansion, radiation, phase changes, molecular behaviour, and effects on measuring devices.

The physical quantity is assigned at the point where underlying dynamics produce an observable effect. This is why manifestation is a useful concept for broadening physics. It directs attention to the transition between what may exist beneath observation and what becomes measurable.

The observable result is not necessarily identical to the underlying process. A shadow is produced by light, obstruction, and geometry. The shadow is real, but it is not a separate material object. It manifests the absence or reduction of light within a particular relation. A wave pattern on water manifests energy transfer through the medium, but the water as a whole does not travel with the wave across the entire surface. A spectrum manifests the interaction between matter and radiation, but the coloured lines are not independent substances hidden inside the atom.

The manifestation carries information about the underlying system. Physics works backward from consequence toward structure. This is the logic of much experimental science. A detector records a trace. A screen shows a pattern. A material changes state. An instrument registers a signal. The scientist asks what physical process could generate that result.

The interpretation is rarely immediate. Several models may produce similar manifestations. Additional experiments are needed to distinguish them. This means observation itself does not arrive already labelled. A detector click is an event. Calling it the detection of a particular particle requires theoretical context, calibration, exclusion of alternatives, and a chain of inference. The measurement is real. The meaning of the measurement is constructed through disciplined reasoning.

This distinction is vital when physics deals with entities that cannot be observed directly. Science often knows them through effects. The interior of a star is inferred through emitted light, oscillations, and theoretical models. Subatomic processes are reconstructed from detector events. Gravitational structure is inferred from motion, timing, lensing, and wave signals. Indirect observation is not weak science. Much of physics depends on it.

But indirect observation requires intellectual honesty about the layers involved:

- the recorded event;
- the physical quantity extracted;
- the model used;
- the entity inferred;
- the interpretation proposed.

Manifestation sits at the beginning of this chain. It is what reality makes available to observation.

In a UED-oriented view, manifestation has an even broader role. Matter, fields, motion, and measurable quantities can be approached as expressions of deeper energy dynamics within SVE. But the word must not become a vague substitute for explanation. To say that something "manifests" is not enough. A scientific account must show how the deeper condition produces the observable consequence, under what constraints, and with what measurable relation.

Manifestation is therefore not an escape from mathematics. It is the domain mathematics must eventually formalize. It reminds physics that the measurable surface may arise from deeper structure.

4. Consequence Before Cause

Physics often asks what caused an event. What caused the object to accelerate? What caused the material to heat? What caused the particle to appear? What caused the star to collapse?

The question is natural, but scientific inquiry usually begins from the opposite side. We first encounter the consequence. The cause is inferred. A body changes motion. A detector records a signal. A material deforms. A current appears. A spectrum changes. These observable consequences motivate the search for explanation.

The cause is rarely seen in isolation. It is reconstructed through relations among events. This is why consequence deserves a central place in physics. A consequence is not merely the final stage of an event. It becomes a new condition for later events.

A collision changes momentum and material structure. That change influences what happens next. Heat transfer alters temperature. The new temperature changes pressure, phase, resistance, or reaction rate. A field variation changes particle motion. The altered distribution modifies the field.

Physical reality is therefore not a sequence of independent causes followed by passive effects. It is a continuous consequence chain. Each event inherits conditions from earlier events and modifies conditions for later ones.

This perspective is especially important in complex systems. No single cause may fully explain the outcome. Weather emerges from interacting gradients, flows, radiation, rotation, moisture, terrain, and prior atmospheric states. A material fracture may arise from load, microscopic defects, fatigue, temperature, geometry, and manufacturing history. The cause is distributed across the system. The consequence results from alignment among several conditions.

Classical physics often simplifies this complexity into a net force, initial state, and equation of motion. This is useful when the system permits it. But the simplification should not be mistaken for the entire causal architecture. A broader physics asks not only which quantity changed, but how the system acquired the conditions that made the change possible.

What gradient developed? What boundary constrained the system? What instability grew? What relation became misaligned? What threshold was crossed? What earlier consequence created the present condition?

These questions shift attention from isolated causes toward evolving structure. They also help distinguish explanation from labelling. Saying that an object moved because of force may identify a valid mechanical relation. But if force is calculated from the motion, the explanation may remain incomplete. Saying that a gas expanded because pressure changed may describe the event without explaining why the pressure changed. The quantity identifies the link. The consequence chain provides the wider process.

Universal Energy Dynamics places strong emphasis on this continuity. Under UED, reality unfolds through consequences generated within a structured energy field. A local change does not remain isolated. It modifies the surrounding conditions, creating new differences and possible responses. The Law of Consequences, in this view, is not a moral principle. It is an attempt to describe continuous physical propagation: every resolved or unresolved delta alters the next state of the system.

For this idea to become physics rather than philosophy, it must be expressed through specific models and testable cases. Still, the conceptual shift is valuable even before full formalization. Physics may be broader than the study of separate quantities because nature does not evolve one variable at a time. It evolves as interconnected consequence.

5. Quantities Are Interfaces, Not Foundations

A physical quantity stands between reality and theory. On one side lies the physical process. On the other lies measurement, mathematics, and interpretation. The quantity provides an interface.

Mass allows motion and gravitational response to be compared. Temperature allows thermal states to be ordered and related. Pressure allows distributed mechanical interaction to be represented. Charge allows electromagnetic behaviour to be classified and calculated. Entropy allows transformations and statistical accessibility to be described.

These quantities are scientifically real because they correspond to stable and reproducible patterns. But an interface is not necessarily the foundation of the system.

A computer screen shows icons, windows, files, and buttons. These are real operational features. They allow the user to control complex processes. Yet the computer is not fundamentally made of icons. The interface compresses deeper operations into a form the user can manage.

Physical quantities may function similarly. They are not visual decorations. They connect reliably with deeper processes. But they present selected aspects of those processes in forms suitable for measurement and calculation.

This analogy should not be pushed too far. Nature is not necessarily a computer, and quantities are not merely fictional icons. The point is structural. An accessible representation can be real and useful without being ontologically fundamental.

Temperature behaves like this. It is a meaningful property of a macroscopic system, yet it arises from deeper statistical dynamics. Pressure behaves similarly. Even solidity can be treated as an emergent interface between human-scale objects, despite the microscopic structure beneath it.

This raises an important possibility: many quantities currently treated as fundamental may later be understood as emergent. History supports caution. Heat was once treated differently before thermodynamic and molecular theories matured. Space and time changed status under relativity. Particles changed status under quantum field theory. Gravity changed from force to geometry.

A quantity can retain all its practical validity while its foundational interpretation changes. This is why the search for the "origin" of a physical quantity can be misleading. The quantity itself may originate in scientific description. What needs physical explanation is the stable pattern that makes the quantity meaningful.

Instead of asking: *Where does force come from?* We may ask: *What structured interaction produces the acceleration or deformation measured as force?*

Instead of asking: *Where does temperature come from?* We ask: *What collective energy distribution produces the measured thermal state?*

Instead of asking: *What is charge made of?* We ask: *What underlying structure produces the stable electromagnetic response described as charge?*

The questions become more precise because they separate vocabulary from manifestation. This does not imply that every quantity must eventually disappear. Some relations may prove foundational within the deepest available theory. But fundamentality should be earned, not assumed.

A quantity becomes fundamental relative to a framework when the theory treats it as irreducible. A later framework may reorganize it. A broader physics therefore uses quantities confidently while keeping their status open. It calculates through the interface. It investigates beneath the interface. This dual movement protects both practical science and foundational inquiry.

6. What Quantitative Physics Can Miss

Measurement requires selection. Selection creates a blind region. When physics isolates a variable, it gains precision but temporarily excludes other aspects of the event. This is usually necessary. To study pendulum motion, colour may be irrelevant. To analyse a circuit, the cultural history of the laboratory is irrelevant. To calculate an orbit, the chemical composition of the planet may often be neglected. Scientific power comes from identifying what can be ignored safely.

But the habit of exclusion can become a worldview. What is not quantified may begin to appear unreal, unscientific, or unimportant. This is dangerous because some physical structures are recognized qualitatively before reliable measurement becomes possible. A new phase of matter may first appear as an unexpected behavioural change. An instability may be visible before its control parameter is identified. A complex system may display organization that cannot be explained by measuring isolated parts alone.

Emergence, coherence, boundary conditions, and system architecture can be physically decisive even when they are not represented by one simple quantity. Consider two arrangements made from the same material and containing the same total mass. Their behaviour may differ completely because their geometry differs. A bridge and a pile of steel are not mechanically equivalent. The quantities of material may match. The organization does not. Structure matters.

Likewise, the same components can produce different outcomes depending on timing, sequence, phase, or alignment. A collection of oscillators may behave independently or synchronize. A material may remain stable until a threshold is crossed. A fluid may move smoothly and then become turbulent. The quantities do not vanish, but the system-level pattern becomes essential.

Physics must therefore study relations among quantities, not only the quantities themselves. It must study how organization alters consequence. The difficulty is that patterns can tempt vague explanation. Words such as coherence, emergence, alignment, complexity, and organization can sound deep while remaining poorly defined. A broader physics must not use them as substitutes for measurement. It must develop measurable indicators of structural behaviour.

The challenge is to expand quantification, not to escape discipline. This is where institutions face a real tension. Science must protect itself from unsupported claims. Requiring operational definitions, repeatability, and mathematical clarity is essential. But if institutional standards become too narrow, research may favour only questions that fit existing instruments and theories. Foundational questions may be dismissed because they are not yet measurable within the current framework.

The balance is difficult. Too much openness allows speculation to masquerade as science. Too much closure allows present methods to define the limits of reality. The solution is not to lower evidential standards. It is to distinguish stages of inquiry. A conceptual proposal is not yet a theory. A theory is not yet confirmed. A pattern is not yet a mechanism. But each stage can be legitimate if its status is stated honestly.

Physics becomes broader when it allows disciplined movement from observation to concept, from concept to model, and from model to test. Quantity remains central. It no longer monopolizes the meaning of science.

7. Toward a Science of Manifestation

If physics is more than quantities, how should it be defined? Calling physics *"the science of matter and energy"* is useful, but it may be too narrow if matter and energy themselves are treated only as finished categories. Calling it *"the science of measurable phenomena"* protects scientific discipline, but it may imply that reality begins only where current instruments can reach.

A broader definition might be: **Physics is the study of the structures, dynamics, relations, and consequences through which reality becomes physically manifest and measurable.**

This definition includes quantities without reducing physics to quantities. It preserves experiment. It preserves mathematics. It also includes emergence, patterns, fields, geometry, system organization, and the transition from deeper dynamics to observable effects.

Such a physics would ask several levels of question:

- At the descriptive level: *What happened?*
- At the quantitative level: *How much changed?*
- At the relational level: *Which variables changed together?*
- At the structural level: *What configuration made the event possible?*
- At the dynamical level: *How did the system evolve?*
- At the foundational level: *What underlying reality gives rise to the observable quantities and fields?*

Mainstream physics already operates across many of these levels. The broader definition does not require abandoning established practice. It makes explicit what physics has always done at its best. Newton did more than calculate. He reorganized motion. Faraday did more than measure. He reconceived space around electrical and magnetic systems. Einstein did more than improve numerical accuracy. He changed the physical meaning of space, time, gravity, and motion. Quantum pioneers did more than introduce new constants. They confronted the breakdown of classical categories.

Physics advances when measurement and conceptual reconstruction work together. Universal Energy Dynamics enters this broader landscape as a proposal that physical manifestations arise through the dynamics of a universal structured substrate—SVE. Its central claim is not that quantities are false. It is that quantities belong to the manifestation layer.

Mass, force, charge, temperature, pressure, and other variables describe how organized physical systems respond within deeper energy dynamics. Matter itself is approached as structured substrate rather than an independent substance placed inside empty space. Fields become distributed modes of substrate condition. Force becomes the measured response to gradients and interactions.

This picture may open useful connections across physical domains. But the scientific burden is clear. A science of manifestation cannot remain only a philosophical language. It must identify the mechanisms connecting substrate dynamics with measurable quantities.

How does SVE organization produce mass-like response? How do field structures arise? How are known equations recovered? What observations distinguish UED from existing models? Where can the proposal fail?

These questions determine whether the doorway leads into physics or remains at its philosophical boundary. The broader scope of physics gives UED permission to ask foundational questions. It does not give UED permission to bypass evidence. That distinction is healthy.

Physics is more than quantities because reality is more than what any current quantity captures. But physics remains science only when deeper claims eventually return to observable consequence.

Closing Reflection: The Number and the Phenomenon

Return to the rainbow. A physicist can measure the angles. Light can be separated into wavelengths. Refraction can be calculated. Droplet size can be studied. The observer's position can be included. The equations may predict where the arc will appear. Every quantitative step deepens understanding.

Yet the rainbow remains a relational event. Remove the droplets, and it disappears. Change the light, and it changes. Move the observer, and the apparent rainbow moves. No single object in the sky contains it. No isolated quantity is the rainbow. It manifests when several physical conditions align.

This is why physics is more than quantities. Quantities describe the conditions. They measure the relations. They predict the consequences. But the phenomenon arises from organization. Reality does not present itself as separate numbers waiting to be collected. It presents structured events, changing systems, gradients, interactions, thresholds, patterns, and consequences. Science develops quantities to grasp these manifestations with precision. The achievement is enormous. But the measurement should not erase the event from which it came.

A number tells us how much. A pattern tells us how things are arranged. A consequence tells us what changed. A theory tells us how selected relations may be organized. A deeper physics asks how all these layers belong together.

This wider view does not make physics vague. It makes physics more complete. It recognizes that measurement is one stage of inquiry rather than the whole of reality. It allows us to ask what exists before a quantity is defined, what structure makes measurement possible, and how the measured result arises from underlying dynamics.

The final hierarchy may therefore be written simply: **Reality → Pattern → Interaction → Consequence → Manifestation → Measurement → Quantity → Equation → Theory → Interpretation**

Physics moves across this entire hierarchy. It does not begin at the bottom with the equation. It begins at the top with reality. The task of science is to travel carefully between the layers without confusing one for another. A theory should not be mistaken for the event. A quantity should not be mistaken for the process. A measurement should not be mistaken for the whole manifestation. And the unknown should not be mistaken for the unreal.

Physics became powerful by learning to count. It will become deeper by remembering what it is counting. The rainbow can be measured. But before it became a set of wavelengths, it was light finding a visible form through a structured relationship. Perhaps all physical quantities stand in a similar position. They are not reality reduced to numbers. They are reality becoming readable.

Chapter 8 — What Lies Beneath Fields?

The Vacuum Chamber

Imagine a chamber from which nearly everything has been removed. The air has been pumped out. Dust has been reduced. No visible object moves inside. The walls are isolated as carefully as technology permits. To ordinary perception, the chamber appears empty.

What remains? Classical intuition answers: almost nothing. Modern physics gives a more complicated reply. Fields do not necessarily disappear when ordinary matter is removed. The electromagnetic field remains part of the theoretical description. Quantum fields are not erased simply because no particles are detected. The gravitational structure of the surrounding world does not cease at the chamber wall. The vacuum itself may possess measurable physical properties.

What looked like emptiness has gradually become one of the most conceptually crowded regions in physics. This reversal captures the history of physical thought. At first, matter seemed primary and space seemed empty. Then force connected distant matter. Fields gave structure to the surrounding region. Relativity made spacetime dynamic. Quantum theory made the vacuum active. The deeper physics moved, the less "nothing" resembled nothing.

Yet the field concept, powerful as it is, creates a new question: *What is a field?*

A field is often described as a physical quantity assigned to every point in space and time. The description is mathematically precise. It allows values, directions, changes, interactions, and propagation to be calculated. But the definition may describe how a field is represented without telling us what physically exists.

Is a field a real continuous entity? Is it a mathematical structure used to organize measurable effects? Is it a state of spacetime? Is it an excitation of something deeper? Does the field exist independently, or only through relations and interactions?

Physics can calculate field behaviour with extraordinary success while these questions remain open at the foundational level. This is not unusual. Force worked before its mechanism was understood. Thermodynamics worked before molecular explanations matured. Quantum mechanics worked before agreement existed about its interpretation. A concept may become operationally secure long before its ontology is settled.

Fields may occupy that position today. They are not speculative decorations. Electromagnetic, gravitational, and quantum field theories have produced some of the most successful descriptions in science. But their success does not automatically prove that fields are the final substance of reality.

Perhaps fields are fundamental. Perhaps they are emergent. Perhaps they are different expressions of one deeper structure. Perhaps geometry, information, and matter all arise together from an underlying physical substrate.

The purpose of asking what lies beneath fields is not to weaken field theory. It is to continue the movement that produced field theory in the first place. Physics moved beneath force because force did not explain enough. It may now need to move beneath fields for the same reason.

1. Fields Solved One Mystery and Inherited Another

The field concept solved a major problem in classical physics. If one object influenced another across distance, how did the effect reach the second object? Direct action through empty space appeared conceptually unsatisfactory. The field offered a local description.

A source altered the physical condition of the surrounding region. Another object responded to the field at its own location. The interaction no longer needed to leap mysteriously across space. This was a profound improvement.

Electromagnetic fields could vary, propagate, carry energy, and produce measurable consequences. Gravitational fields organized motion. Later, quantum fields provided the formal setting from which particles and interactions could be described. The field became one of the central ideas of modern physics.

Yet a concept that resolves one mystery may inherit another. If an object responds to a field, what is the field itself? Saying that a field is a quantity distributed through space tells us how to represent it. It does not necessarily tell us what occupies the region.

Suppose a magnetic field exists around a magnet. The field can be mapped through the response of a compass, charged particles, or other instruments. Its strength and direction can be calculated. Changes in the field can produce electrical effects. These consequences establish that the field description corresponds to something physically significant.

But the instrument never captures "field substance" in isolation. It responds through interaction. We know the field through what it does. This resembles the earlier status of force. Force was inferred from acceleration and deformation. Field is inferred from local response and propagation. The description has moved deeper, but the logical structure remains relational.

Quantum field theory intensifies the issue. Within its formalism, particles are associated with excitations of fields. This gives fields greater ontological priority than particles. Yet quantum fields do not behave like ordinary continuous materials. They are described through mathematical structures that include probabilities, operators, symmetries, and quantized excitations.

The word *field* may therefore create a misleading familiarity. A classical field can be pictured as a continuous value spread through space, much like temperature distributed through a room. A quantum field is not simply a classical substance with small vibrations added to it. Its formal and conceptual structure is far more demanding. The same word covers different theoretical layers.

Physics also contains several fields rather than one obvious universal field. Different particle types and interactions are associated with different field structures. This creates a further question. Are these genuinely separate foundational entities? Or are they different modes of a deeper common reality?

Unification has repeatedly succeeded in physics. Electricity and magnetism became electromagnetism. Space and time became spacetime. Mass and energy entered a deeper relation. Several apparently different particle interactions were reorganized within shared theoretical structures. Every successful unification encourages the suspicion that present multiplicity may not be final.

But suspicion is not proof. A deeper theory cannot be accepted merely because unity is aesthetically satisfying. The field concept has earned its place through prediction and experiment. Any proposal beneath fields must explain why field theories work so well, preserve their successful results, and produce additional consequences.

The challenge is not to replace fields with a vaguer word. It is to understand whether fields are the foundation or the visible architecture of something beneath them.

2. The Unfinished Meeting of Gravity and Quantum Physics

One of the clearest signs that present physics may not be complete is the difficulty of bringing general relativity and quantum theory into one fully unified framework. Both theories are extraordinarily successful.

General relativity describes gravity through spacetime geometry. It works across planets, stars, gravitational lensing, cosmology, and many large-scale systems. Quantum theory describes microscopic behaviour with exceptional accuracy. Quantum field theory organizes particles and non-gravitational interactions through fields, symmetries, and probabilistic structures. Each theory has survived demanding tests.

Yet their foundational languages differ. General relativity treats spacetime as a dynamic geometric structure. Quantum field theory usually formulates fields with respect to a spacetime background, even when that background may be curved. In one framework, geometry is part of the physical dynamics. In the other, quantum states and fields carry the central description.

The difficulty becomes severe under conditions where both strong gravity and quantum effects matter—such as the earliest physical conditions considered in cosmology, the deep interior of black-hole models, or extremely small scales. The equations that work brilliantly in their own domains do not combine easily into one complete description. This is not merely a technical inconvenience. It may indicate that one or both frameworks are effective descriptions arising from a deeper level.

Einstein spent much of his later career searching for a unified field theory. He hoped that gravitation and electromagnetism might be expressions of a common structure. His efforts did not produce the accepted unification he sought. The episode is instructive. Even a physicist who transformed the meaning of gravity did not consider the existing separation of physical interactions intellectually final.

Later approaches have pursued quantum gravity through different routes. Some begin from quantized geometry. Others begin from extended objects, networks, information, holographic relations, or new mathematical structures. The variety of approaches shows both creativity and uncertainty. No single picture has yet become the universally accepted physical foundation from which all relevant phenomena clearly follow.

This does not mean modern physics has no direction. It means the boundary is genuine. Black holes sharpen the problem. General relativity describes a horizon and extreme gravitational structure. Quantum theory raises questions about radiation, information, and the evolution of states. The resulting information problem is not simply a disagreement over words. It exposes tension among deep principles: gravitational geometry, quantum evolution, thermodynamics, and the meaning of information.

When several successful theories meet and produce contradiction, physics cannot solve the problem by declaring one vocabulary sacred. It must investigate the layer from which the conflicting descriptions may emerge. Historically, such conflicts have been productive. The mismatch between classical mechanics and electromagnetism helped motivate relativity. The failure of classical physics at atomic scales helped produce quantum theory.

The present difficulty between quantum theory and gravity may similarly indicate that spacetime, fields, and particles are not independent foundations. Perhaps they arise together. Perhaps geometry is quantum. Perhaps fields are geometric. Perhaps both are manifestations of a deeper relational structure. The search continues because the present theories are too successful to discard and too difficult to combine without transformation. That is usually where physics becomes most interesting.

3. Is Information More Fundamental Than Matter?

Information has become increasingly important in modern physics. It appears in statistical mechanics, quantum theory, black-hole discussions, computation, thermodynamics, and the study of complex systems. Some physicists and philosophers have therefore proposed that information may be more fundamental than matter.

The proposal is attractive. Physical systems can be distinguished by states. States can encode differences. Interactions can transfer, preserve, erase, or reorganize those differences. Quantum theory places strong constraints on what can be known, copied, or measured. Black-hole physics raises questions about whether information can be lost. The success of information language suggests that reality may be understood not only through objects and energy, but through distinguishable structure.

Yet the word *information* can easily become another reified abstraction. Information is always information about a difference, state, relation, pattern, or possibility. A written sentence carries information because physical symbols differ in an organized way. A digital bit requires a physical distinction between possible states. A genetic sequence carries information through molecular arrangement. A quantum state is described informationally because different preparations produce different measurable probabilities.

The informational description depends on physical distinction. This raises a basic question: *Can information exist without a physical structure capable of carrying difference?* If not, then information may be fundamental as a relation, but not as a free-floating substance. Saying "everything is information" may therefore explain less than it appears to. It may replace matter with a word that still requires a physical basis.

Information also depends on a level of description. A gas can be described microscopically through the states of many particles or macroscopically through temperature and pressure. The amount and type of information depend on which distinctions the observer considers relevant. This does not make information subjective in a trivial sense. Physical constraints determine which states are possible and how they evolve. But information always involves structure and distinguishability.

The idea becomes especially powerful when connected to entropy. Entropy can be interpreted through the number of microscopic configurations compatible with a macroscopic condition. Information and thermodynamic accessibility become related. This has led to deep insights connecting computation, energy cost, and physical systems.

Still, the informational description does not eliminate the physical system. A memory device must be built from something. A signal must be embodied. A state must exist in a field, material, geometry, or other physical structure. The stronger possibility is therefore not that information replaces matter, but that matter and information are different descriptions of structured difference.

Matter presents stability. Information presents distinguishability. Energy presents capacity for transformation. Fields present distributed condition. Geometry presents relation. These may be different conceptual views of one deeper physical organization.

In a substrate-based framework, information would not float above reality. It would arise from distinguishable configurations within the substrate. A stable difference in SVE organization could carry information because it changes the possible consequences of later interaction. This interpretation preserves the importance of information while refusing to treat it as magic. Information matters because physical reality can take different structured forms. Without form, relation, or contrast, there is nothing to inform.

4. Is Geometry the Foundation—or a Manifestation?

Einstein's theory of gravity elevated geometry from a descriptive tool to a central part of physical reality. Before relativity, space and time were often treated as a fixed stage. Matter moved within them. Geometry described the stage but did not necessarily participate in physical events.

General relativity changed that picture. Spacetime could curve. Its geometry depended on matter and energy. The paths of objects and light reflected that geometry. The stage had become dynamic. This was a major conceptual advance because it replaced gravitational force with structure. A planet moved as it did not because an invisible agent continually pulled it, but because the spacetime through which it moved had a particular geometry. The explanation was deeper.

But does this make geometry the final foundation? Geometry is a language of relation. It expresses distance, curvature, direction, connection, and dimensional structure. It can describe how events are arranged and how paths evolve. Yet geometry itself may be represented mathematically in many possible forms. Not every possible geometry is physically realized. The question remains: *What makes one geometry the geometry of the physical universe?*

General relativity connects geometry with matter-energy through field equations. But the theory does not necessarily explain why spacetime exists at all or whether it remains fundamental at every scale. Quantum gravity research often challenges the idea of smooth continuous spacetime at the smallest scales. Some approaches suggest that spacetime may emerge from more basic relations, networks, quantum states, or patterns of entanglement.

If so, geometry would occupy a position similar to temperature. Temperature is real and measurable at the macroscopic level, yet it emerges from microscopic dynamics. Spacetime geometry might be real at large scales while emerging from deeper non-geometric structure. This would be another example of physics moving beneath its own successful vocabulary.

The possibility is difficult because all ordinary physical description already assumes spatial and temporal relations. To imagine something "beneath spacetime" is almost impossible using everyday language. Words such as *before*, *inside*, *outside*, and *beneath* are themselves spatial or temporal. The conceptual tools may be built from the very structure being questioned.

This is one reason mathematics becomes essential. It can express relations that ordinary visualization cannot. But the mathematics still requires physical interpretation. An emergent geometry must show how familiar spacetime behaviour arises, why relativity works, and what measurable deviations may appear under extreme conditions.

A substrate view faces a similar challenge. If SVE is described as existing "in space," then space remains more fundamental. If SVE is said to produce spacetime, then its own structure cannot be explained using ordinary spatial language without care. The substrate would need relational organization from which measurable space and time emerge.

This pushes UED beyond a simple medium model. SVE cannot be merely a modern version of mechanical ether filling a pre-existing container. It must be proposed as the structured basis from which matter, fields, and possibly geometry itself become manifest. That is a much stronger claim. It is also a much heavier scientific burden. Geometry may be the deepest established description of gravity. But the question of whether geometry itself is fundamental remains open.

5. Emergence: When the Whole Becomes Physically Real

Emergence occurs when a system displays stable properties or behaviour that are not useful to describe at the level of isolated components.

Temperature emerges from collective microscopic dynamics. Pressure emerges from distributed interactions. A wave emerges from coordinated change across a medium or field. A crystal lattice emerges from many constituents settling into an ordered structure. Turbulence emerges from fluid motion under particular conditions.

In each case, the higher-level property is real. It produces measurable consequences. But it is not carried fully by any single component. Emergence helps explain how physics can contain several valid descriptive layers.

A molecule does not possess the temperature of an entire gas. Yet the gas genuinely has a temperature. An individual water molecule is not wet. A large collection of water molecules can produce wetness through interaction with surfaces. A single atom is not elastic in the same way a material beam is elastic. Elasticity arises from the organized response of many atoms within a structure.

This means "not fundamental" does not mean "unreal." It means dependent on organization. The concept of emergence may apply more widely than once assumed. Particles may emerge from fields. Classical objects may emerge from quantum behaviour. Spacetime may emerge from deeper relations. Fields themselves may emerge from a substrate.

The strength of emergence is that it allows physics to preserve different levels without forcing one vocabulary to describe everything. The danger is that emergence can become a placeholder. Saying that something "emerges" does not explain how it emerges. A scientific theory must identify the components, interactions, thresholds, and conditions through which the higher-level pattern appears.

For temperature, statistical mechanics provides a detailed bridge. For many complex systems, the bridge is harder. For spacetime or quantum fields, it remains an open research challenge. Emergence also complicates reductionism. Reductionism seeks to explain wholes through parts. This approach has been extraordinarily successful. But if the behaviour of the whole depends on organization, boundary conditions, and collective relations, listing the parts may not be enough.

The same components arranged differently can produce completely different outcomes. A functioning engine and a pile of engine parts contain similar materials. Their physical capacities differ because structure differs. Organization carries causal importance.

This insight aligns strongly with UED. If matter is structured SVE, then physical properties arise not only from *what* exists but from *how* energy is organized, aligned, constrained, and stabilized. Mass, charge, field behaviour, and material form could then be approached as emergent consequences of substrate configuration.

Again, this cannot remain a poetic statement. A serious model must specify which structures produce which properties. It must show why stable forms persist. It must explain how one configuration differs from another. It must derive measurable quantities from deeper organization.

Still, emergence offers a conceptual path beneath fields. Fields need not be rejected. They can remain physically real at their level while arising from a more fundamental structure. The deeper substrate would not compete with fields in the same way one object competes with another. It would explain how field-like behaviour becomes possible.

This layered view may be more faithful to the history of physics than the demand for one final object. Reality may not be built from a single smallest thing. It may be built through levels of structured manifestation.

6. Why the Search Continues

Physics continues searching beneath fields because important questions remain unresolved.

Why do the known fields exist? Why do they possess their particular symmetries and constants? Why do particles have the properties they do? Why does spacetime have its observed structure? Why is gravity so different in strength and formulation from other interactions? Why do quantum theory and general relativity resist complete unification? What is the physical nature of dark matter? What drives the large-scale accelerated expansion described through dark-energy models? Why does the universe contain more matter than antimatter in the observed form? What determines the transition from quantum possibility to classical outcome?

These questions differ in status. Some are direct empirical puzzles. Some are theoretical tensions. Some are questions of interpretation.

They should not be mixed carelessly. But together they show that the current framework, successful as it is, has not closed foundational inquiry.

The institutional structure of physics reflects this tension. Large experiments search for predicted particles, rare interactions, gravitational signals, dark-matter candidates, and deviations from established models. Theoretical programmes explore quantum gravity, unification, cosmology, information, and emergent spacetime. Some approaches receive substantial support. Others remain marginal. This is unavoidable to some extent. Resources are limited. Institutions must choose which questions appear promising, testable, and connected to existing evidence.

But institutional momentum can also shape what counts as a legitimate question. A framework with a large research community develops sophisticated mathematics, experimental proposals, journals, and training pathways. An alternative approach may struggle not only because it is wrong, but because it lacks comparable institutional geometry.

This does not justify accepting outsider theories without scrutiny. Most broad proposals fail because they are vague, mathematically incomplete, or experimentally empty. Institutional resistance often protects science from unsupported claims. Yet history also shows that conceptual change can begin outside the dominant framework or at its uncomfortable edge.

The proper standard should therefore be neither authority nor rebellion. It should be consequence. Does the proposal explain known results? Does it solve a real problem? Does it generate a distinctive prediction? Can it be tested? Can it be wrong?

The search beneath fields continues because no current term should be mistaken for a final answer simply because it occupies the deepest layer of today's equations. Force once seemed fundamental. Then fields became deeper. Spacetime geometry became deeper for gravity.

Quantum fields became deeper for particles. Information and emergence now challenge the apparent independence of those structures.

The search is not evidence that physics knows nothing. It is evidence that physics knows enough to see where its knowledge stops. That boundary is not a weakness. It is the active frontier.

7. The Possibility of a Deeper Substrate

A substrate is a foundational physical basis through which observable structures and dynamics become possible. Historically, substrate ideas have often taken the form of a medium. Sound travelled through air. Water waves travelled through water. It therefore seemed natural to assume that light must travel through an ether.

The classical ether proposals eventually failed to achieve the required experimental and theoretical status. Relativity removed the need for a mechanical light-carrying medium of the older kind. This history makes physicists understandably cautious about new substrate proposals.

A modern substrate cannot simply repeat the old ether with a different name. It cannot be an invisible substance added whenever explanation becomes difficult. It must not contradict relativity by introducing an undetected preferred mechanical frame without consequence. It must not use the word *energy* vaguely. It must explain how fields, particles, geometry, and measurable quantities emerge. It must produce testable distinctions.

Universal Energy Dynamics proposes Structured Vacuum Energy as such a foundational basis. The phrase contains three important claims:

- **Universal** means that the substrate is not confined to particular objects or locations.
- **Energy** means that the foundation is dynamically capable of

change, transmission, organization, and consequence.

- **Structured** means the vacuum is not treated as featureless emptiness. It possesses organization sufficient to support fields, matter forms, gradients, and stable patterns.

Within this view, matter is not inserted into the substrate as a foreign substance. Matter is a stable structured form of it. Fields are not separate invisible additions surrounding matter. They are extended conditions or modes of substrate organization. Force is not a primary agent. It is the measurable response of matter forms to gradients, differences, and changing substrate conditions. Spacetime may represent the relational geometry of substrate dynamics rather than an independent container. Information may arise from distinguishable substrate configurations.

This picture offers conceptual unity. But unity is not enough. The central scientific questions are demanding.

What variables describe SVE? How is its structure represented mathematically? How do known field equations emerge? How does SVE preserve relativistic consistency? What produces quantization? How do stable particles or matter forms arise? How are mass and charge derived? What new phenomenon does UED predict? What observation could falsify it?

Without answers, SVE remains a philosophical framework. That does not make it worthless. Philosophy can identify hidden assumptions and organize new questions. But it should not be presented as completed physics. The responsible position is neither premature certainty nor premature dismissal.

A deeper substrate is possible because current physics already treats the vacuum as physically non-trivial, fields as fundamental to present theory, and spacetime as dynamic. The conceptual door is open. What remains is to determine whether SVE provides a better physical architecture than existing approaches. The burden belongs to the theory. The reader should not be asked to accept the substrate. The reader should be invited to see why the question arises.

Perhaps fields are the deepest reality. Perhaps fields and geometry emerge together. Perhaps information captures the underlying structure. Perhaps a universal substrate generates all three. The correct answer will not be decided by vocabulary. It will be decided by explanatory power, mathematical coherence, and measurable consequence.

Closing Reflection: The Emptiness That Would Not Stay Empty

Return to the vacuum chamber. Nearly all ordinary matter has been removed. Yet complete emptiness remains difficult to define. Fields still belong to the description. Geometry still structures relations. Quantum theory refuses to treat the vacuum as simple nothingness.

The chamber has become a test of our language. When we say *empty*, do we mean no air? No particles? No measurable excitations? No fields? No spacetime? No structure? Each answer belongs to a different level.

The history of physics has repeatedly transformed emptiness into hidden activity. Space first appeared passive. Fields made it active. Relativity made it dynamic. Quantum theory made the vacuum physically consequential. The next step may reveal another layer.

But we should resist the temptation to fill every unknown with a preferred story. Information may be fundamental—or it may describe structured difference. Geometry may be fundamental—or it may emerge. Quantum fields may be the final foundation—or they may be effective expressions of deeper dynamics. A substrate may unify these layers—or it may become another unsupported medium.

The correct scientific attitude is not to choose the most impressive word. It is to keep the hierarchy clear. We observe manifestations. We measure quantities. We construct theories. We test consequences. We interpret the structures. We identify what remains unresolved. Only then do we propose a deeper layer.

The search beneath fields continues because fields answer the question of how physical influence is distributed, but not necessarily the final question of what reality is. They describe local conditions with immense precision. They generate particles, interactions, waves, and measurable effects within modern theory. Yet the existence, multiplicity, and structure of the fields themselves still invite explanation.

A deeper physics may eventually show that fields, geometry, matter, and information are not separate foundations. They may be different manifestations of one organized reality. Universal Energy Dynamics names that possibility *Structured Vacuum Energy*. The name is not the proof. The picture is not the theory. The theory is not established until it returns to measurement.

But every new physical framework begins by seeing an old category differently. Force became field. Gravity became geometry. Particles became excitations. Vacuum became structure. The next question is therefore unavoidable: *What makes structure possible?*

The vacuum chamber appears empty. Physics has learned not to trust that appearance. Perhaps the deepest reality is not hidden behind the things we see. Perhaps it is the structured condition that allows anything to appear at all.

Chapter 9 — Toward a New View of Reality

What Is the Table Made Of?

Place your hand on a table. It feels solid. It supports your weight. It resists pressure. It occupies space. It appears to possess a clear boundary separating it from the air around it. Ordinary experience tells us that the table is made of matter.

Physics gives a more complex answer. The table consists of molecules and atoms. Atoms contain internal structures. What appears as continuous solidity arises from organized interactions at scales far below direct perception. The resistance felt by the hand is not evidence that matter is a perfectly packed, indivisible substance. It is a macroscopic consequence of deeper physical organization.

The table is solid. But solidity is not the whole story of what the table is. This distinction has appeared repeatedly throughout the book. Force is real as a measurable relation, yet it may arise from fields, gradients, geometry, and material response. Temperature is real at the level of a system, yet no single particle possesses the temperature of the whole. Pressure is measurable, yet it emerges from collective interaction. Particles appear as localized events, yet quantum field theory places fields beneath the classical particle picture.

Physical reality repeatedly reveals the same pattern: what appears fundamental at one scale may be a stable manifestation of something deeper. This raises an unavoidable question. If matter is not ultimately a collection of tiny solid objects, what makes material form possible? If fields are more fundamental than particles, what gives fields their structure? If the vacuum is not simple emptiness, what kind of physical foundation does it represent?

Universal Energy Dynamics begins from the possibility that the universe is not made of objects placed inside an empty container. It proposes instead that reality possesses a continuous, structured foundation through which matter, fields, motion, and observable physical effects arise. This foundation is called Structured Vacuum Energy, or SVE.

The name should not be treated as proof. Naming a deeper layer does not establish that the layer exists exactly as described. A useful scientific proposal must eventually be expressed mathematically, connected with established physics, and exposed to experiment. But a concept can still be valuable before it becomes a complete theory. It can reorganize the question.

Instead of asking how separate objects communicate across emptiness, we may ask how different forms of one underlying structure interact. Instead of treating matter and fields as unrelated categories, we may ask whether both are modes of the same physical foundation. Instead of treating force as an independent agent, we may see it as the measurable response of structured matter to changing energetic conditions.

The table beneath your hand may then be understood not as a block of independent substance, but as a stable form within a deeper, continuous reality. Its solidity is genuine. Its separateness may be less absolute than it appears.

1. Why Another View Is Needed

Modern physics is extraordinarily successful. It can describe planetary motion, electromagnetic behaviour, atomic structure, radiation, thermodynamic processes, and the large-scale evolution of the observable universe. Its predictions support technologies that would have appeared impossible only a few centuries ago.

A new view of reality should therefore not begin by declaring established physics useless. That would be intellectually careless. The need for a broader view arises not from the failure of physics to calculate, but from the incompleteness of its conceptual unity.

Different areas of physics use different foundational languages. Classical mechanics speaks through bodies, mass, momentum, and force. Thermodynamics speaks through temperature, entropy, pressure, and equilibrium. Relativity speaks through spacetime geometry. Quantum theory speaks through states, probabilities, operators, and measurement. Quantum field theory speaks through fields, symmetries, interactions, and excitations. Each language works powerfully within its domain.

But the relationship among these domains remains conceptually difficult. Is matter fundamentally made of particles? Are particles excitations of fields? Are fields fundamental entities? Is spacetime a background, a dynamic geometry, or an emergent structure? Is information more basic than matter? Does energy exist as a substance, a conserved quantity, or a property of physical configuration?

Physics often possesses several accurate maps without one universally accepted account of the common territory. This is not unusual in science. Before electromagnetism, electric and magnetic phenomena were studied separately. Before statistical mechanics, thermodynamics described heat without fully connecting it to microscopic behaviour. Before relativity, space and time appeared independent and absolute.

A deeper theory often emerges when several successful descriptions are recognized as manifestations of one broader structure. The search for unity is therefore reasonable. But unity can be pursued badly. One may invent a universal substance and claim that it explains everything merely because every phenomenon is assigned to it. Such a theory becomes impossible to test because every outcome is interpreted as confirmation.

A genuine foundational proposal must do more. It must reduce conceptual fragmentation. It must preserve successful existing results. It must define its terms clearly. It must distinguish description from ontology. It must eventually produce testable consequences.

SVE is introduced within this discipline. It is not proposed because current physics knows nothing. It is proposed because current physics knows enough to expose several unfinished boundaries. The vacuum possesses physical significance. Fields extend throughout regions without ordinary matter. Particles arise within field descriptions. Matter and energy are deeply connected. Spacetime is dynamic. The old worldview of solid objects moving through absolute emptiness has already been abandoned.

SVE asks whether these developments point toward a common physical basis. The proposal is not that every accepted equation should be discarded. It is that the equations may describe different manifestation layers of one structured reality.

The need for another view therefore comes from a simple question: *Can matter, fields, geometry, energy, and observable quantities be understood as different expressions of one continuous physical foundation?* That question does not oppose modern physics. It grows from its direction of travel.

2. The Vacuum Is Not Simply Nothing

The word *vacuum* usually suggests absence. An empty bottle contains less matter than a full one. A vacuum chamber has much of its air removed. Interplanetary space appears almost empty compared with Earth's atmosphere. In ordinary language, vacuum means that things are missing.

Physics complicates this meaning. Removing ordinary matter does not necessarily remove physical structure. Gravitational relations remain. Electromagnetic fields may remain. Quantum field theory does not treat the vacuum merely as a blank region in which absolutely nothing exists. The vacuum is described as the lowest-energy state of fields, not the absence of all physical possibility.

This distinction is important. A lowest-energy state is still a state. It possesses structure within the theory. The vacuum can be connected with fluctuations, field behaviour, and measurable consequences. It is not simply philosophical nothingness.

However, care is required. The statement that the vacuum has physical properties does not automatically prove any specific substrate theory. Modern field theory can describe vacuum effects without assuming a mechanical medium resembling air, water, or classical ether. One must not take the non-emptiness of the quantum vacuum and treat it as direct confirmation of SVE. It only opens the conceptual door.

Historically, physicists often assumed that waves required a material medium. Since light behaved like a wave, they proposed a luminiferous ether filling space. The ether seemed reasonable because known waves required something to wave. Yet the classical mechanical models of ether did not achieve lasting experimental support. Relativity reorganized the problem without needing the old medium.

This history provides an essential warning. A substrate cannot be introduced merely to satisfy intuition. It must not be a hidden material added beneath every unexplained phenomenon.

SVE must therefore differ from the old ether in several ways. It should not be imagined as a gas, fluid, or elastic solid filling pre-existing space. Such an image would simply transfer familiar matter into an invisible background. Instead, SVE is proposed as the foundational physical condition through which space, matter, fields, and energetic distinctions become possible.

This is a more radical claim. The substrate does not sit inside the universe. It is part of what makes the physical universe possible. The word *structured* is therefore crucial. A completely featureless foundation could not produce differentiated phenomena unless some mechanism of distinction were added.

For matter, fields, and motion to arise, the underlying reality must be capable of variation, organization, persistence, and transformation. It must support difference. It must permit local states to influence surrounding states. It must allow patterns to stabilize. It must allow disturbances to propagate.

This does not mean SVE should be pictured as a visible material. The human mind prefers images, but every image may mislead. Perhaps SVE is closer to a continuous relational architecture than to a substance in the everyday sense. Perhaps what we call space is one manifestation of its organization. Perhaps what we call energy is its capacity for changing configuration. Perhaps fields express distributed states within it.

At present, these remain interpretive possibilities. The scientific task is to turn them into precise relations. Still, one conceptual shift can already be made: vacuum need not mean nothing. It may mean the absence of ordinary matter within a deeper physical structure that remains active.

3. Structured Vacuum Energy as a Foundational Proposal

Structured Vacuum Energy can be introduced through three ideas.

The first is **continuity**. Reality is not approached as a collection of absolutely separate objects placed within emptiness. Physical forms remain connected through the foundation from which they arise.

The second is **structure**. The foundation is capable of differentiated organization. It can support regions, gradients, patterns, oscillations, alignments, and stable configurations.

The third is **energy**. The foundation is dynamic. Its configurations can change, interact, propagate consequences, and form persistent structures.

These ideas provide a conceptual architecture. They do not yet constitute a complete mathematical theory. The word *energy* itself must also be used carefully. In established physics, energy is a measurable conserved quantity associated with the state and evolution of systems. It is not always treated as a substance.

When UED speaks of Structured Vacuum Energy, the term should not suggest a hidden fuel filling space. It refers more broadly to the dynamic capacity of the foundational structure to form, change, and transmit consequence. This usage must eventually be connected rigorously with established definitions of energy. Otherwise, the word risks becoming too elastic. A scientific term that can mean everything explains nothing.

The central value of SVE lies in the unity it attempts to provide. Matter would no longer be one kind of reality and fields another. Both would arise from the organization of the same foundation. Space would no longer be an empty container independent of physical content. It would express relational structure within the foundation. Force would no longer need to act as an invisible object. It would quantify the response of stable forms to gradients and changing conditions. Particles would not be tiny foreign objects travelling through SVE. They would be localized, persistent patterns within it.

This picture resembles other historical moves in physics. Heat was reinterpreted through microscopic motion. Sound was understood as organized disturbance through a medium. Particles were reinterpreted through quantum fields. Gravity was reinterpreted through geometry. Each shift placed the familiar phenomenon within a deeper process.

SVE proposes a similar shift at the foundational level. But it must avoid claiming more than it has earned. A universal substrate sounds explanatory because everything can be assigned to it. The real test is whether the assignment produces new understanding.

Can it explain why different matter forms possess different properties? Can it connect gravity and quantum behaviour? Can it account for the stability of particles? Can it recover known field equations? Can it clarify the role of vacuum structure? Can it predict an effect not already contained in existing theories?

Until such questions receive formal answers, SVE should be treated as a structured research proposal. It provides a lens. The lens may reveal connections. But the lens must be tested against the physical world.

The strongest introduction to SVE is therefore modest: SVE is the proposed continuous and structured physical foundation from which matter, fields, and observable energy dynamics may emerge. The words *proposed* and *may* are not signs of weakness. They mark the proper scientific status of the idea.

4. Matter as Stable Form

Matter appears stable because its patterns persist. A stone remains recognizably the same stone over time. A table retains its shape. A planet maintains a coherent form. An atom preserves characteristic behaviour.

Yet none of these forms is completely static. Atoms vibrate. Molecules interact. Materials exchange energy with their surroundings. Planets move. Stars burn. Even structures that appear motionless participate in continuous physical processes.

Stability therefore does not mean absence of dynamics. It means persistence through dynamics. This distinction is central to a substrate view. Matter need not be imagined as frozen substance. It can be understood as organized activity that maintains a stable configuration.

A whirlpool offers a simple analogy. The whirlpool has a recognizable shape and location. It can persist for some time. Yet the water composing it changes continuously. The whirlpool is not separate from the water. It is a stable pattern within the water's motion.

Matter under SVE would not be identical to a whirlpool, and the analogy should not be taken literally. But it illustrates how form can persist without being made from permanently fixed contents. A flame provides another analogy. Its shape may remain relatively stable while fuel, gas, heat, and radiation continually pass through the process. The flame is real. Its reality lies in organized continuity rather than static substance.

Similarly, a physical particle may be approached as a localized stable mode of an underlying field or substrate. Its identity would arise from the persistence of a pattern, not from an indivisible material bead. This view aligns with the direction of quantum field theory, where particles are associated with excitations of fields.

SVE extends the question further: could the fields themselves be structured modes of one deeper foundation? If so, matter would be a stable form of stable field behaviour arising within the substrate. This layered emergence would help explain why matter can appear solid while being deeply dynamic.

Material properties would depend on organization. Mass could express resistance associated with stable energetic structure. Charge could express a characteristic mode of relational interaction. Spin and other quantum properties could arise from the geometry or dynamics of the configuration.

Again, these are research directions, not completed derivations. The concept becomes scientifically meaningful only when particular structures can be related to particular measured properties. Still, understanding matter as stable form produces several useful changes.

It weakens the opposition between matter and energy. Matter becomes organized energy dynamics. It weakens the opposition between object and field. An object becomes a localized field or substrate configuration. It weakens the idea of absolute separateness. Forms remain distinct enough to interact, yet they arise within a continuous foundation.

It also changes how persistence is understood. A thing remains "the same" not because nothing changes, but because its organization continues to resolve internal and external disturbances without losing identity. Stability becomes dynamic coherence.

This may be a more realistic picture of the physical world. A stable form survives because its internal relations maintain a recognizable pattern. When those relations can no longer be sustained, the form transforms, decays, breaks, melts, reacts, or reorganizes. Matter is then not the opposite of change. Matter is change held within a persistent geometry.

5. Energy Dynamics and Manifestation

If matter is stable form, physical events can be understood as changes in form, relation, or distribution. Energy dynamics refers to this continuous activity.

A system is heated. A material expands. A charge redistributes. A wave propagates. A body accelerates. A chemical bond changes. A star collapses. Each event involves a difference between earlier and later conditions. The difference may exist in position, concentration, pressure, temperature, field strength, geometry, or internal organization.

Systems respond to these differences. The response produces manifestation. A wave becomes visible on water. A material glows. A detector registers a signal. A body changes motion. A field alters the behaviour of matter. Manifestation is therefore the observable face of underlying dynamics.

This framework places consequence at the centre. A local change modifies surrounding conditions. Those changed conditions produce further responses. Physical reality evolves through linked consequences. This is more continuous than the picture of separate objects delivering isolated forces.

Consider sound. A vibrating source changes the local state of the surrounding medium. Nearby regions respond. The disturbance propagates. Eventually, a biological ear or instrument converts the pattern into another form. No single air molecule travels from the speaker to the listener. The manifestation arises through local transmission.

The same general logic may apply across deeper physical layers. A change in one region of SVE could alter nearby conditions. The altered structure could propagate as a ripple, field variation, or wave-like response. Stable matter forms would interact with this changing structure according to their own configuration.

What classical physics describes as force could emerge as the measured change in motion. What field theory describes as field interaction could reflect distributed substrate dynamics. What quantum theory describes probabilistically could arise from the constrained possibilities of ripple alignment and resolution.

This broader interpretation remains speculative until formalized. But it has conceptual value because it provides one continuous sequence:

Difference → Response → Propagation → Interaction → Manifestation → Measurement

The sequence avoids treating physical quantities as agents. Energy does not "want" to move. Fields do not "decide" to interact. Systems respond because their conditions are structured and non-uniform.

The language of dynamics should also avoid implying that all systems simply move toward one universal state. Physical reality contains stable cycles, oscillations, growth, decay, feedback, resonance, and self-organization. The same foundation can support many forms of consequence.

A gradient may dissipate. A disturbance may oscillate. A structure may stabilize. Two patterns may interfere. A threshold may be crossed. The visible world emerges from these different modes of resolution.

In SVE, manifestation is not added after the dynamics. It is the point at which the dynamics becomes locally observable through matter, fields, radiation, geometry, or instrument response. This helps explain why physical quantities belong to the manifestation layer. They measure what the dynamics has produced. They do not necessarily name the foundation from which it arose.

6. Physical Quantities as Observer Vocabulary

A universe without observers would still contain physical processes. Stars would form. Matter would interact. Waves would propagate. Structures would emerge and dissolve. But there would be no kilograms, metres, joules, kelvins, coulombs, or pascals.

These units belong to measurement. Measurement belongs to the relationship between an observed system, an instrument, a standard, and an interpreting intelligence. This does not make physical quantities subjective fantasies. Nature constrains them. A bridge cannot carry any load simply because an engineer chooses a convenient number. A thermometer cannot display any value and remain accurate. A theory cannot predict anything and still claim correspondence with reality.

Physical quantities are objective within disciplined measurement. But their objectivity is relational. They translate physical manifestation into observer vocabulary.

The word *observer* must also be used carefully. In physics, an observer need not be a conscious person staring at an event. It may refer to a reference frame, measuring apparatus, or defined system of registration. Yet conscious intelligence is ultimately required to develop the concept, interpret the reading, and construct the theory. An instrument produces a physical response. Science turns that response into meaning.

This distinction becomes especially important in a substrate view. SVE would exist independently of our labels. Matter forms would persist without being assigned mass. Fields would vary without being given numerical strength. Physical changes would occur without being called forces. Quantities enter when observers decide how to compare manifestations.

Mass describes one stable aspect of material response. Charge describes one pattern of electromagnetic participation. Temperature describes the thermal condition of a system. Pressure describes distributed interaction. Entropy describes statistical and thermodynamic accessibility. Force describes change in motion or deformation under interaction.

These quantities are not false. They are partial translations. They convert different expressions of one physical reality into operational terms suited to particular questions. This view explains why several quantities can describe the same event. A falling body can be analysed through force, acceleration, momentum, energy, spacetime geometry, or field behaviour. The event remains one. The vocabularies differ. Each vocabulary selects a relation.

This does not mean all descriptions are equally good. Some are more accurate. Some are more efficient. Some apply only within limited conditions. Some reveal deeper unity. The observer vocabulary must continue to answer to reality. But no one vocabulary should be mistaken for reality itself.

In a UED interpretation, physical quantities could be understood as projection variables: measurable expressions of deeper substrate dynamics as they appear through particular matter forms and experimental arrangements. The word *projection* here should not imply illusion. A shadow is a projection, yet it carries real information about light and obstruction. A graph is a projection of data, yet it reveals structure. A physical quantity similarly projects one aspect of a deeper process into a measurable relation.

The quantity is real at the level of description. The substrate is proposed at the level of ontology. The bridge between them remains the central scientific task.

7. A New View Must Still Become Physics

A new view of reality can be intellectually attractive long before it becomes scientifically persuasive. SVE offers several appealing possibilities. It provides continuity beneath matter and fields. It treats vacuum as structured rather than empty. It interprets matter as stable form. It places force at the level of response. It connects physical quantities with manifestation and measurement. It encourages a unified picture rather than isolated categories.

But intellectual attractiveness can be dangerous. A theory that feels comprehensive may be accepted emotionally before it is tested physically. This book has repeatedly warned against that move. The warning must apply most strongly here.

UED cannot criticize mainstream physics for promoting interpretation beyond evidence and then ask readers to accept SVE without evidence. The same standards must apply to every framework. Several tasks remain essential.

First, definitions must become precise. What exactly is meant by SVE? What variables describe its state? What distinguishes structure from mere metaphor?

Second, known physics must be recovered. A deeper theory should explain why classical mechanics, electromagnetism, relativity, thermodynamics, and quantum predictions work in their established domains.

Third, formal relations must be developed. Conceptual diagrams are not enough. The theory must specify how substrate changes generate measurable outcomes.

Fourth, distinctive predictions are required. A framework becomes physically meaningful when it predicts something different, more precise, or more coherent than existing theories.

Fifth, the proposal must remain falsifiable. There must be possible observations that could show the model to be wrong or incomplete.

These requirements do not weaken UED. They protect it from becoming mythology. A concept can begin philosophically and later become mathematical. Many scientific revolutions began with a change in interpretation before the full formal structure was complete. Faraday's field intuition preceded Maxwell's equations. Einstein's equivalence reasoning preceded the mature form of general relativity. But the later mathematics and experimental tests were essential.

The same path is required for SVE. The current chapter therefore does not present a final theory of reality. It presents a disciplined possibility.

The universe may not be a collection of independent things moving through nothing. It may be a continuous structured foundation expressing itself through stable forms, distributed fields, changing geometries, and measurable consequences. Matter may be organized persistence. Force may be response. Energy may be dynamic capacity. Physical quantities may be observer vocabulary.

If these ideas can be formalized and tested, they may broaden physics. If they cannot, they should remain philosophical propositions. That boundary should not be feared. It is where serious scientific work begins.

Closing Reflection: The Table Revisited

Place your hand on the table once more. It still feels solid. Nothing in this chapter requires you to deny that experience. The table supports objects. It resists pressure. It has measurable mass, temperature, volume, elasticity, and strength. Its physical reality is not threatened by a deeper interpretation.

But the meaning of solidity has changed. The table is no longer imagined simply as a block of self-existing substance placed inside empty space. It can be understood as a stable material form arising from deeper physical organization. Its apparent stillness contains continuous dynamics. Its boundary is operationally real, yet the table remains connected with its surroundings through gravitational, electromagnetic, thermal, and mechanical relations. Its properties are not arbitrary labels. They are measurable manifestations of how the structure responds.

Structured Vacuum Energy offers one possible foundation for this picture. It proposes that the table, the air, the fields around them, and the space through which they relate are not completely separate categories. They may be different expressions of one continuous physical reality.

Within this view:

- Matter is stable structure.
- Fields are distributed structure.
- Motion is changing relation.
- Force is measured response.
- Energy dynamics is the continuous evolution of physical configuration.
- Quantities are the vocabulary through which observers describe the manifestations.

This is a broader picture than the world of isolated objects. It is also more demanding. A unified idea must explain more, not less. It must produce precision rather than hide uncertainty. It must distinguish metaphor from mechanism. It must return from philosophy to experiment.

SVE should therefore be held neither as unquestionable truth nor as an idea to be dismissed merely because it is unfamiliar. It should be treated as a doorway. The doorway opens toward a physics in which emptiness is structured, matter is dynamic, and measurement describes manifestation rather than exhausting reality. Whether the doorway leads to a deeper physical theory will depend on what can be built beyond it.

For now, the table remains beneath your hand. Its solidity is still real. But perhaps solidity is not the final fact. Perhaps it is what a deeper reality feels like when its structure becomes stable enough to support you.

Chapter 10 — Physics and the Future

The Student Who Asked One Question Too Many

A student solves a numerical problem correctly. The given values are substituted. The formula is rearranged. Units are checked. The final answer matches the textbook. The class is ready to move ahead.

Then the student asks: *"But what is actually happening?"*

The question may appear unnecessary. The calculation is complete. The required answer has already been found. In an examination, no additional explanation may be needed. Yet this is precisely where physics begins again.

The numerical result tells us *how much*. It may tell us how fast, how far, how hot, how strongly, or with what probability. But the student is asking something else. What changed within the system? Why did matter respond in this way? What physical structure made the response possible? Does the quantity name the cause—or only measure the consequence? Is the equation describing reality, or only one layer of it?

Physics has advanced because certain people refused to stop after the correct calculation. Newton could describe planetary motion, but gravity's mechanism remained open. Faraday saw structure in the region around magnets and charges. Maxwell recognized a deeper unity beneath electricity, magnetism, and light. Einstein questioned assumptions about space, time, acceleration, and gravity that had appeared settled. The pioneers of quantum theory confronted experimental patterns that classical concepts could not comfortably hold.

Their progress did not come from rejecting measurement. It came from asking what the measurement meant. The future of physics may depend on preserving this same movement.

The science has become more mathematically powerful, technologically capable, and institutionally specialized than at any earlier time. Instruments detect events at scales inaccessible to ordinary perception. Computers process quantities no individual mind could follow. Large collaborations examine signals buried beneath enormous backgrounds of noise. This progress is real.

But greater precision does not eliminate foundational questions. It often reveals them more clearly. The deeper physics looks, the less nature resembles a collection of separate objects carrying fixed properties through empty space. Matter becomes field excitation. Force becomes gradient, interaction, or geometry. Vacuum becomes physically structured. Space and time become dynamic. Information, emergence, and measurement enter discussions once dominated by objects and motion.

Physics appears to be moving toward a broader account of manifestation. The future may therefore require two forms of discipline at once: the discipline to measure with increasing precision, and the discipline to ask what measurement represents. Physics must remain quantitative. But it may no longer be sufficient to define it only as quantitative science.

The student's question remains open: *What is actually happening?*

1. The Quantitative Revolution Is Not Finished

Any attempt to broaden physics must first respect what quantitative science has achieved. Measurement transformed curiosity into reproducible knowledge. Before reliable measurement, observations remained difficult to compare. One person's "fast" could be another's "moderate." One region's unit of length could differ from another's. Heat, pressure, motion, and weight could be experienced without being expressed through shared standards.

Quantification changed this. It made observations portable. A result obtained in one laboratory could be checked in another. An engineering drawing could guide construction far from the designer. Astronomical predictions could be compared with future events. Experimental deviations could be identified rather than absorbed into vague language.

Mathematics then connected the quantities. The result was not merely more accurate description. It was predictive power. The same equations that described known behaviour could reveal unknown consequences. A theory could be tested by events that had not yet occurred. Technology could be designed before physical construction began.

The quantitative revolution is therefore not something physics should move beyond in the sense of leaving it behind. It is unfinished because measurement continues to expand. New instruments reveal new scales. More precise clocks test time under different gravitational and motional conditions. Detectors record particles and interactions that cannot be perceived directly. Telescopes examine radiation arriving from distant regions and earlier cosmic conditions. Interferometers detect distortions far smaller than ordinary mechanical experience could imagine.

Every improvement in measurement creates a new boundary between known and unknown. A small discrepancy can expose a hidden effect. A repeated deviation can challenge a theory. A previously invisible pattern can become a new physical domain.

The future of physics will therefore remain deeply quantitative. It will require better sensors, more accurate models, stronger mathematics, greater computational capacity, and improved control over experimental conditions.

But quantitative expansion creates a risk. As datasets become larger and models become more complex, understanding can be mistaken for calculation. A system may predict accurately while remaining conceptually opaque. An algorithm may identify a pattern without explaining why the pattern exists. A simulation may reproduce behaviour without revealing which features of the model correspond to physical reality.

This problem will intensify as artificial intelligence becomes more involved in scientific work. Machine-learning systems can search parameter spaces, classify signals, optimize experiments, and detect correlations beyond unaided human ability. They may even generate mathematical structures that fit observations remarkably well.

But prediction alone does not settle meaning. A model may tell us that an event is likely. Physics must still ask what process produces it. The quantitative future therefore needs conceptual intelligence more than ever. The role of the scientist cannot be reduced to feeding data into increasingly powerful systems. Someone must decide what counts as a physically meaningful variable, which assumptions are hidden in the model, what alternative explanations remain, and what experiment could distinguish them.

Numbers do not interpret themselves. The quantitative revolution succeeds when precision and understanding remain connected. Its future weakness would not be too much mathematics. It would be mathematics separated from physical meaning.

2. From Quantitative Science to Manifestation Science

The phrase *manifestation science* does not propose a replacement for quantitative physics. It proposes a wider frame.

Quantitative science asks:

- How much?
- How fast?

- How strong?
- How frequently?
- With what probability?
- Under what measurable conditions?

Manifestation science adds another set of questions:

- What is becoming observable?
- What deeper structure produces this measurable effect?
- How does one physical layer appear through another?
- Why does this quantity arise at this scale?
- What remains hidden when the event is compressed into a number?

These questions already exist within physics. Thermodynamics studies temperature and entropy as macroscopic manifestations of collective behaviour. Condensed-matter physics studies how new properties emerge from the organization of many constituents. Particle physics reconstructs invisible processes from detector signals. Cosmology infers large-scale structure from radiation, motion, lensing, and statistical distribution. Quantum theory relates preparation and measurement to probabilities of observable outcomes.

The broader perspective is therefore not foreign to science. It makes explicit something physics already does. A manifestation is not a decorative appearance placed on top of reality. It is the form through which underlying physical structure becomes locally observable.

Consider a spectrum. The pattern of lines is visible or detectable. But the lines are not independent objects hidden inside the atom. They manifest transitions and structured relations within matter and radiation.

Consider pressure. The gauge reading is real. Yet it condenses countless interactions into one measurable value.

Consider gravitational lensing. The distorted image is observable. It manifests a relation among light, matter, geometry, and the observer.

A manifestation-centred physics pays attention to these transitions between levels. It asks how underlying organization becomes measurable consequence. This approach may be especially valuable where several descriptions coexist. A system can be represented through force, energy, momentum, field, geometry, probability, or information. Each vocabulary captures a different aspect.

Manifestation science does not immediately choose one as final. It asks how they are connected. What is described as force at one level may appear as a field gradient at another. What is described as a particle may appear as a field excitation. What is described as temperature may appear as a statistical property of many microscopic states. What is described as spacetime geometry may eventually emerge from deeper relations.

The concept therefore encourages *vertical physics*: not only relations within one level, but transitions between levels. How does microscopic dynamics become macroscopic behaviour? How does a distributed field produce localized detection? How does a continuous process generate apparently discrete outcomes? How does stable matter arise from underlying activity? How do observer-defined quantities relate to observer-independent dynamics?

These are difficult questions because they cross disciplinary boundaries. They require mathematics, experiment, philosophy, computation, and careful language. But this broader scope may help physics avoid two extremes. The first is reduction without emergence—the belief that naming the smallest known constituent automatically explains the larger system. The second is emergence without mechanism—the use of words such as complexity, coherence, or manifestation without showing how the higher-level behaviour arises.

A mature manifestation science must connect the layers formally. It must show how quantities emerge, not merely state that they do. The future of physics may therefore be broader without becoming less precise. It may move from measuring isolated variables toward explaining the architecture through which measurable variables become possible.

3. The Questions That Remain Open

The future of physics is not difficult to justify. Its open questions are substantial.

Quantum theory and general relativity remain extraordinarily successful within their domains, yet their complete foundational unification remains unresolved. The nature of dark matter is still unknown. Its gravitational consequences are inferred across astronomical systems, but its physical constitution has not been conclusively established. The accelerated expansion of the universe is described through dark-energy models, yet the underlying physical meaning remains unsettled.

The relationship between quantum measurement and definite outcomes continues to generate multiple interpretations. The physical status of the wavefunction remains debated. The emergence of classical reality from quantum descriptions is understood operationally in important ways, but foundational questions persist.

The origin of the observed asymmetry between matter and antimatter remains an active problem. The values of fundamental constants are measured, but why they possess those particular values is not known. The structure of spacetime at extremely small scales remains uncertain. The nature of time itself—whether fundamental, emergent, relational, or thermodynamic—continues to challenge physics and philosophy. Even the concept of vacuum remains complicated.

These questions differ in evidential status. Dark matter is associated with strong observational patterns requiring explanation. Particular dark-matter candidates remain hypothetical. Cosmic expansion is observed through specific data and models. Its ultimate cause remains interpreted. Quantum probabilities are experimentally confirmed. The ontology attached to them remains contested.

This distinction must remain clear. The phrase *open question* should not imply that everything is equally uncertain. Physics possesses islands of extraordinary reliability surrounded by deeper conceptual boundaries. The mistake is to confuse the shoreline with the absence of land. The presence of unresolved questions does not weaken established achievement. It identifies where new work is needed.

Institutions face a difficult task in deciding which questions to pursue. Some experiments require decades, large teams, and enormous resources. Funding a project means not funding another. Theoretical communities develop around promising frameworks, but no institution can explore every possibility equally. This creates unavoidable scientific politics. Dominant programmes gain continuity, tools, and authority. Alternative approaches may remain underdeveloped.

But the existence of institutional bias does not mean every excluded idea is revolutionary. Most speculative proposals fail because they lack precision, evidence, or testable consequence. The future requires a better boundary between openness and discipline. New proposals should not be dismissed simply because they challenge current ontology. Nor should they be accepted merely because they claim to explain what mainstream physics has not yet resolved.

A serious foundational proposal must answer clear questions:

- What established problem does it address?
- What existing results does it preserve?
- What concepts does it define more clearly?

- What formal structure does it introduce?
- What prediction differs from current theory?
- What observation could show it to be wrong?

The unanswered questions of physics are therefore not invitations to invent freely. They are invitations to construct responsibly. The future will belong neither to rigid orthodoxy nor to unrestricted speculation. It will belong to frameworks that connect conceptual depth with measurable consequence.

4. Wonder Is Not the Opposite of Rigour

Science often begins in wonder. A child sees the Moon and asks why it does not fall. A student watches light bend through glass and asks what light is. An observer sees a magnet act without contact and asks what occupies the space between. Wonder begins where familiarity breaks. It notices that the obvious is not actually obvious.

But scientific education can gradually weaken wonder. Students learn definitions before encountering the mystery that produced them. They memorize the inverse-square law without feeling the conceptual difficulty of action across distance. They calculate temperature without asking how collective behaviour becomes a single variable. They learn that particles are field excitations without pausing over what a field is. The completed answer replaces the original question.

This is efficient for examination. It is dangerous for foundational thinking. Wonder is sometimes treated as emotional or unscientific, while rigour is associated with removing curiosity in favour of formal procedure. This is a false opposition. Wonder generates the question. Rigour disciplines the answer. Without wonder, science becomes technical repetition. Without rigour, wonder becomes mythology.

The strongest physicists have often held both. Faraday's experimental imagination did not weaken his seriousness. Einstein's thought experiments were acts of disciplined wonder. Schrödinger's cat was not a whimsical story added to quantum theory—it was designed to expose a conceptual difficulty. Feynman's curiosity did not prevent mathematical precision. It helped him see familiar problems from different angles.

Wonder matters because it prevents language from becoming invisible. A student who continues asking what force actually means may notice that the concept has changed across theories. A researcher who asks what the vacuum really is may refuse to treat a mathematical ground state as a completed ontology. A physicist who asks why information must be physically embodied may avoid turning information into a mystical substance. Wonder protects science from premature closure.

But wonder must also accept correction. Nature is under no obligation to match human intuition. The universe may be stranger than our preferred pictures. A beautiful conceptual idea may fail experimentally. An ugly approximation may work. A familiar analogy may become misleading at deeper scales. Mature wonder therefore does not ask reality to confirm imagination. It allows reality to transform imagination.

This is especially important for UED. The idea of a universal structured substrate can evoke powerful intellectual satisfaction. It seems to restore continuity, connect matter with fields, and place quantities within a larger architecture. That sense of coherence can motivate research. It cannot substitute for research.

Wonder opens the SVE question: *Could matter and fields arise from one structured foundation?* Rigour asks: *What equations follow? What observations distinguish the framework? What error would disprove it?*

Both voices are necessary. The future of physics depends on preserving the child who asks why and developing the scientist who accepts that the answer must survive reality.

5. The Future Scientist Will Need More Than Calculation

The future physicist will likely work in a scientific environment very different from that of Newton, Faraday, or Einstein. Experiments are becoming more complex. Datasets are becoming larger. Research is increasingly collaborative. Mathematical tools are more specialized. Artificial intelligence will assist in modelling, pattern detection, literature analysis, experimental design, and perhaps even formal derivation.

This will increase scientific capability. It may also fragment understanding. A researcher may become highly skilled in one computational or experimental task while remaining distant from the conceptual architecture of the wider problem. Large collaborations need specialization. No individual can master every instrument, equation, and dataset involved in modern high-energy physics, cosmology, or quantum technology.

But foundational questions require synthesis. Someone must ask how the parts fit together. The future scientist therefore needs at least four forms of literacy.

- **Quantitative literacy.** Without mathematics, measurement, and error analysis, physical claims cannot be disciplined adequately.
- **Conceptual literacy.** The scientist must understand what the quantities represent, what assumptions the model makes, and where one theory gives way to another.
- **Historical literacy.** Scientific concepts are easier to question responsibly when one knows how they developed. History reveals that current language was not inevitable. Force, field, atom, space, time, and vacuum all changed meaning.
- **Philosophical literacy.** This does not mean turning laboratories into metaphysical debate clubs. It means

recognizing distinctions between observation, inference, model, interpretation, and ontology. These distinctions protect physics from both dogma and speculation.

Education must therefore change if physics is to broaden. The classroom should not present equations as commandments. It should show where they came from, what they assume, where they work, and what questions they leave open. Students should learn not only to solve a problem but to identify the level of description. Is this a mechanical model? An energy representation? A field description? A statistical approximation? A geometric account? A quantum probability? What has been omitted? Why is that omission justified?

This kind of teaching does not reduce examination performance. It gives calculation meaning.

The future scientist must also learn intellectual status-labeling. A confirmed measurement should be called a measurement. A model should be called a model. An interpretation should be called an interpretation. A speculation should be called a speculation.

This simple discipline may become increasingly important as scientific communication accelerates. Complex ideas are often compressed into dramatic public claims. Mathematical possibility becomes headline reality. Preliminary evidence becomes discovery. Interpretive language becomes fact.

The future of physics requires scientists who can resist this pressure. Precision is not only numerical. It is also linguistic. A theory may contain twelve decimal places of mathematical accuracy and still be discussed conceptually with careless certainty. The next generation will need to be precise in both ways.

6. The Place of UED in the Journey Ahead

Universal Energy Dynamics enters this book near the end for a reason. The reader first had to see how physical quantities arise, how force changes meaning, how fields replaced direct action, how mathematics represents rather than becomes reality, and why deeper questions remain open. Only then can SVE be introduced without appearing as an unsupported answer placed above existing physics.

UED proposes a broad possibility: reality may possess a continuous structured foundation. Matter may be a stable organization within it. Fields may be distributed modes of its condition. Motion may express changing relation. Force may be a manifestation variable describing response. Physical quantities may be observer-developed vocabulary for recurring consequences.

This picture is conceptually coherent. It gathers several themes developed throughout the book. But coherence is the first stage, not the final one. The future of UED depends on whether it can move through a demanding sequence:

concept → definition → model → mathematics → prediction → experiment → revision

Many theories remain trapped at the first two stages. They offer a persuasive vocabulary but no operational bridge. A mature UED must not merely say that SVE causes everything. It must show how particular SVE structures produce particular measurable effects. It must identify which existing equations emerge as approximations. It must state where current theories remain complete enough and where UED predicts a difference. It must expose itself to failure.

The future value of UED may lie not only in providing a new physical model, but in restoring a neglected hierarchy to physics:

- Reality precedes description.
- Dynamics precedes quantities.
- Manifestation precedes measurement.
- Measurement precedes theory.
- Theory precedes interpretation.

This hierarchy can improve physical thinking even before every UED claim is formalized. It prevents force from becoming a substance merely because it appears in an equation. It prevents fields from becoming final simply because they are currently fundamental in a theory. It prevents mathematics from becoming ontology merely because it predicts well. It prevents the vacuum from being called empty merely because ordinary matter has been removed.

UED should therefore proceed neither defensively nor triumphantly. It does not need to declare that all of modern physics is mistaken. Nor should it demand institutional recognition before meeting institutional standards. Its strongest path is constructive: clarify one problem at a time, develop one measurable relation at a time, compare one prediction at a time. Allow criticism to refine the framework. Preserve what works. Discard what fails.

A foundational theory should not fear correction. If its purpose is to describe reality, reality must be allowed to reshape it. The journey ahead is therefore open. UED may develop into a formal substrate theory. It may contribute only selected concepts. It may require major revision. Some parts may fail while others survive. That uncertainty is not embarrassing. It is scientific.

The important point is that the door has been opened responsibly. Physics may be more than quantities. The next step is to show exactly how much more.

7. A Broader Physics for a Broader Reality

The future of physics may not arrive as one final equation replacing everything before it. It may arrive as a better understanding of how several valid layers connect. Classical mechanics may remain the correct language for machines and ordinary motion. Thermodynamics may remain indispensable for macroscopic systems. Quantum theory may remain necessary for microscopic behaviour. Relativity may remain the deepest established account of gravitation and spacetime.

A future substrate theory may not erase these frameworks. It may explain why they arise in different conditions. This would be less like replacing one map with another and more like understanding how several maps belong to the same terrain. A road map, geological map, and weather map are not rivals. They describe different structures. The future theory would need to show how the layers relate without flattening them into one vocabulary.

This is why manifestation may become a central concept. Manifestation allows different levels to be real without requiring each level to be fundamental. Solidity can be real at the human scale. Temperature can be real at the thermodynamic scale. Particles can be real as localized detections. Fields can be real as distributed physical structures. Spacetime can be real as measurable geometry. A deeper substrate can still underlie them. Reality may therefore be layered without being divided.

This broader physics would study not only laws within levels, but transitions between them. How does substrate structure become field? How does field become stable matter? How does quantum behaviour become classical appearance? How does microscopic distribution become thermodynamic state? How does physical interaction become measurable quantity? How does measurement become scientific knowledge?

These questions connect physics with information, complexity, computation, philosophy of science, and perhaps eventually the study of observation itself. But the expansion must be controlled. A broader physics should not absorb every philosophical question and call it physical. Its boundary remains measurable consequence. What broadens is not the permission to speculate endlessly. What broadens is the recognition that measurement arises within a deeper architecture deserving investigation.

The future may also require greater humility about finality. Every major physical framework once appeared capable of completing the picture. Classical mechanics inspired dreams of a fully determined universe. Electromagnetism suggested deep unity. Relativity transformed space and time. Quantum theory overturned classical assumptions. The Standard Model organized an extraordinary range of particles and interactions. Yet none has ended the search.

Perhaps there will be a final theory. Perhaps reality supports an endless hierarchy of effective descriptions. Perhaps the question of *finality* itself reflects the human desire for closure more than the structure of nature. Physics need not decide this in advance. Its task is simpler and harder: follow the consequences, improve the descriptions, test the interpretations, keep the unknown visible.

A broader reality requires a physics confident enough to calculate and humble enough to question what the calculation means.

Closing Reflection: The Question After the Answer

Return to the student in the classroom. The numerical answer is still correct. Nothing in this book has made the formula less useful. The object still accelerates. The bridge still carries load. The circuit still obeys measurable relations. The satellite still follows a calculable path. The quantitative achievement of physics remains intact.

But another layer has appeared. The formula now stands inside a larger hierarchy. It describes a relationship among quantities. The quantities describe measurable manifestations. The manifestations arise through physical interaction and structure. The structure may emerge from deeper fields, geometry, or substrate dynamics. The answer is correct. It may not be complete.

This distinction is the future opening of physics. The next revolution may not begin by rejecting equations. It may begin by asking where equations stand. It may not begin by discovering another quantity. It may begin by explaining why quantities arise. It may not begin by adding another particle. It may begin by understanding matter as stable organization. It may not begin by adding another force. It may begin by seeing force as response within structured reality. It may not begin by declaring fields final. It may begin by asking what makes field-like behaviour possible.

Physics has already travelled a long distance. It moved from objects to forces. From forces to fields. From fields to geometry. From particles to quantum excitations. From empty space to active vacuum. From deterministic trajectories to probabilities. From isolated properties to information and emergence.

The next step cannot be predicted with certainty. But the direction is visible. Physics is moving toward deeper structure and broader relation. Universal Energy Dynamics names one possible direction within that movement. It suggests that matter, fields, motion, and physical quantities may arise through a continuous structured energetic foundation.

Whether SVE becomes accepted physics will depend not on confidence, authority, or elegance, but on what it can explain, calculate, and predict. That is how it should be. The future of physics must remain open to imagination and answerable to evidence. Wonder must ask the question. Mathematics must discipline the proposal. Experiment must test the consequence. Reality must give the final response.

Perhaps physics will one day explain how every known quantity arises from a deeper structure. Perhaps it will discover that fields and spacetime are manifestations of something more fundamental. Perhaps new instruments will expose patterns that current theories cannot anticipate. Perhaps the deepest progress will come not from finding smaller objects, but from understanding relation, organization, and consequence.

Whatever the future brings, one principle should remain: a correct answer should never be used to silence a deeper question.

The student asks again: *"But what is actually happening?"* This time, the teacher does not rush to the next formula. The teacher pauses.

Because that question is not outside physics. It is where the future of physics begins.

Final Reflection

Reality Is Deeper Than Its Vocabulary

Physics gives names to what nature does. It names motion, force, energy, momentum, pressure, temperature, charge, field, entropy, space, and time. It assigns symbols to them. It defines units. It connects them through equations. It builds theories capable of predicting outcomes with extraordinary precision.

Through this vocabulary, humanity has learned to describe falling bodies, planetary motion, electricity, magnetism, heat, light, atoms, stars, and the large-scale structure of the universe. This achievement should not be reduced. It is one of the greatest expressions of human intelligence.

But every vocabulary creates a danger. Once a word becomes familiar, we may begin to think that the word is the thing. Once a quantity becomes measurable, we may assume that the quantity is the reality. Once an equation predicts correctly, we may believe that the equation has revealed everything that exists beneath the event.

The language becomes so successful that it disappears from our attention. We stop seeing it as language. We begin seeing it as nature itself.

A stone falls, and we say gravity acts. A spring returns, and we say restoring force acts. Heat moves, and we say energy flows. A gas expands, and we say pressure drives the motion. An electron responds, and we say a field acts upon it.

These statements are useful. They allow calculation. They connect events to established theories. But the words do not perform the events.

Gravity does not read an equation. Energy does not knowingly travel. Pressure does not make decisions. Fields do not possess intention. Nature does not speak the vocabulary of physics. Physics speaks about nature.

That distinction has guided this book from the beginning. The phenomenon comes first. The quantity comes later. Reality changes. An observer notices the change. A pattern is recognized. A measurable difference is isolated. A quantity is introduced. An equation expresses a relation. A theory organizes the equations. An interpretation attempts to explain what the theory means.

Each layer is valuable. Each layer also has limits. The danger begins when we reverse the order and imagine that reality is generated by the description. Physical laws do not command the universe. They summarize regularities found within it. Equations do not push planets, bend light, heat metals, or move charges. They represent relationships that human intelligence has discovered and formalized. Theories do not contain reality. They contain structured descriptions of selected manifestations of reality.

This does not make physics less real. It makes our understanding of physics more mature. A map is not weakened because it is not the territory. Its value lies in its correspondence with the territory. A map can guide movement, reveal distance, show structure, and prevent error. But no one expects it to contain the trees, rocks, weather, sound, soil, or history of the landscape.

Scientific theories function in a similar way. Newtonian mechanics maps one region of physical behaviour with extraordinary effectiveness. Relativity maps another layer. Quantum theory reveals structures that classical intuition cannot hold. Thermodynamics describes collective behaviour through quantities that do not exist in the same way at the level of individual particles. Each map is powerful. None should be confused with the whole territory.

This is why older theories often survive inside newer ones. Newton does not disappear when Einstein arrives. Thermodynamics does not become useless when statistical mechanics explains temperature and pressure more deeply. Classical objects do not become imaginary because quantum physics reveals deeper structure. A description can remain valid at one scale even when it is no longer fundamental at another.

Reality appears to be layered. The table is solid at the scale of the hand. It is molecular at another scale. It is atomic at another. It is described through quantum fields at a deeper level. Its temperature, mass, strength, and pressure response are all real. Yet none alone is the table. The table is a stable physical form whose different manifestations become visible under different modes of observation.

This may be one of the most important lessons of physics. Reality does not have to choose one vocabulary merely because human beings prefer final answers. It can be described through particles and fields, energy and geometry, statistics and dynamics, depending on the question, scale, and conditions.

The task of deeper physics is not merely to collect more quantities. It is to understand how the layers relate. How does a field become a localized event? How does microscopic behaviour become macroscopic stability? How does matter acquire the properties measured as mass, charge, temperature, or pressure? How does geometry shape motion? How does structure become manifestation? How does manifestation become measurement?

These questions carry physics beyond numerical technique without carrying it outside science. They ask what the quantities represent. They ask what makes measurement possible. They ask why the physical world produces patterns stable enough to be expressed mathematically.

This book has described physical quantities as the manifestation vocabulary of science. That phrase should not be understood as a rejection of quantities. A vocabulary can be precise, objective, and indispensable. Force remains necessary in mechanics. Energy remains central across physics. Temperature, pressure, charge, momentum, and entropy continue to describe real and measurable aspects of physical systems.

But their scientific reality does not automatically make them independent substances. They may express how deeper structures respond, organize, and become observable. Force may describe a change in motion rather than an invisible object acting upon matter. Temperature may describe a collective thermal condition rather than a substance contained inside a body. Pressure may describe distributed interaction rather than a separate agent. Charge may name a stable pattern of electromagnetic participation whose deeper origin remains open. Entropy may express the structure of possible configurations without becoming a universal spirit of disorder.

The quantity is real as a relation. The underlying process may be deeper.

Perhaps information is fundamental. Perhaps these are different manifestations of a deeper structure. Perhaps the future will reveal concepts for which present physics has no vocabulary. We should not decide in advance. Science progresses best when confidence and openness remain together.

We should trust established theories where they work. We should preserve their equations, evidence, and practical achievements. We should also keep their boundaries visible. A theory can be successful without being final. A model can predict without exhausting explanation. A quantity can be real without being fundamental. An interpretation can be reasonable without being measured fact. A speculation can be valuable without being presented as confirmation.

These distinctions do not slow science. They protect its movement. The future of physics may depend not only on more powerful machines or more complicated mathematics, but on clearer thinking about the relationship between reality and representation. It may depend on scientists willing to ask what familiar words conceal. It may depend on teachers who present formulas as achievements rather than commandments. It may depend on students who continue asking what is actually happening after the numerical answer has been found.

Physics began before anyone called it physics. It was present in falling rain, flowing water, fire, shadows, sound, motion, resistance, and the night sky. Human beings encountered consequences first. They developed language later. That language became increasingly precise until it could guide machines, describe atoms, and examine distant galaxies. But the language remains inside the reality it describes.

The universe is not made of our definitions. It does not divide itself into textbook chapters. It does not separate mechanics from thermodynamics, fields from matter, or geometry from energy according to academic departments. Reality is continuous. Our descriptions are selective. The purpose of deeper inquiry is not to erase those descriptions. It is to understand how they belong together.

Perhaps that is what it means to say that physics is more than we know. It does not mean that everything established is wrong. It means that every established answer stands within a reality larger than itself. A force equation may be correct. A field theory may be precise. A geometric model may predict accurately. A quantum calculation may agree with experiment. And still the deeper question may remain.

Reality is not diminished by our vocabulary. Our vocabulary is strengthened when it remembers reality.

The final lesson is therefore simple: Use the quantity. Trust the measurement. Respect the equation. Test the theory. Mark the interpretation. Question the ontology. Remain open to the layer beneath.

Because physics is not merely the language through which we describe the universe. It is the continuing effort to understand why the universe can be described at all. And reality, whatever its deepest nature may be, will always remain greater than the words we use to reach it.

Ten Questions

Every Science Lover Should Ask

Science advances through answers, but it remains alive through questions. A good scientific question does not reject established knowledge. It asks what the knowledge explains, what it assumes, where it applies, and what remains unresolved. It distinguishes observation from interpretation and precision from finality.

The following questions are not intended to produce quick answers. They are meant to deepen the reader's relationship with physics.

1. What Was Actually Observed?

Every scientific claim should begin here. What did the instrument register? What event occurred? What pattern appeared? What changed?

A detector may record a signal. A telescope may collect radiation. A thermometer may respond to a thermal condition. A particle track may appear in a chamber. These are observations. But the meaning assigned to the observation usually requires additional steps.

The signal may be interpreted as evidence of a particle. The motion may be attributed to a field. The pattern may support one theory more strongly than another. These interpretations can be reasonable and scientifically powerful, but they are not identical to the observation itself.

A careful science lover should always ask: *What was directly observed, and what was inferred from it?* This question does not weaken scientific conclusions. It makes their structure visible.

2. What Does the Quantity Really Describe?

Mass, force, energy, temperature, charge, pressure, and entropy are central to physics. But what kind of reality does each quantity possess?

Does the quantity name a substance? A property? A relation? A statistical condition? A measurable response?

Temperature, for example, describes a system-level thermal condition. It is real and measurable, but it does not exist inside a single molecule in the same way it exists for the whole system. Pressure describes distributed interaction. Force describes changes in motion or deformation under particular conditions.

A physical quantity may be indispensable without being fundamental. The important question is not only: *What is its numerical value?* It is also: *What physical manifestation has been compressed into this quantity?*

3. Does the Equation Explain the Event—or Only Describe It?

An equation may predict an outcome with extraordinary accuracy. But prediction and explanation are not always the same.

Newtonian gravity predicts planetary motion. General relativity gives a deeper geometric account. Quantum equations predict measurement outcomes, while their interpretation remains debated.

A formula can show how variables are related without revealing why that relationship exists at a deeper level. When an equation works, we should respect its power. But we should also ask: *What does the equation explain? What does it assume? What does it leave unanswered?*

A correct equation is a major scientific achievement. It need not be the final ontology of the event.

4. What Has Been Idealized or Ignored?

Every useful model leaves something out. A falling body may be treated without air resistance. A pendulum may be modelled with a massless string and frictionless pivot. A fluid may be treated as continuous even though it consists of molecules. A planet may be approximated as a point mass.

These simplifications are not dishonest. They are often necessary. The real question is whether the omitted details matter under the conditions being studied.

A good science lover asks: *What assumptions make this model possible? What has been neglected? Under what conditions would those neglected factors become important?* A model is strongest when its limitations are known.

5. Is the Scientific Term Being Treated as a Thing?

Scientific language often converts relations into nouns. Force acts. Energy flows. Entropy increases. Fields carry information. Particles exchange something.

These expressions are useful, but they can make abstractions sound like independent objects. This is called *reification*. The concept becomes grammatically solid before its physical meaning is fully understood.

Whenever a scientific term appears to act like an agent, we should pause. Is the term naming a physical entity? Or is it summarizing a relation, response, distribution, or pattern?

The question is especially important in foundational physics, where language can quietly become ontology.

6. Could the Same Event Be Described at Another Level?

A physical event often permits several valid descriptions. A falling stone can be described through force, energy, momentum, fields, or spacetime geometry. A gas can be described through temperature and pressure or through microscopic molecular behaviour. A solid can be described as a macroscopic object, an atomic lattice, or a quantum system.

These descriptions need not compete. They may belong to different scales. A science lover should ask: *Which level is being used? Why is it appropriate? What deeper or higher-level description also exists?*

This question protects us from assuming that one successful vocabulary is the only possible account of reality.

7. What Is Established, and What Is Interpreted?

Modern physics contains both highly confirmed results and open interpretations. Quantum interference is experimentally established. The meaning of the wavefunction remains debated. The expansion of the universe is observationally supported. The ultimate physical nature of dark energy remains uncertain.

Gravitational effects attributed to dark matter are strongly observed. The identity of dark matter is not yet settled.

The distinction between established result and interpretive framework is essential. A science lover should ask: *Which part of this claim is measured? Which part is model-dependent? Which part remains speculative?*

Speculation is not the enemy of science. Unmarked speculation is.

8. What Would Show the Theory to Be Wrong?

A scientific theory gains strength by exposing itself to possible failure. A framework that explains every possible outcome after the event is difficult to test meaningfully.

A serious theory should identify consequences that distinguish it from alternatives. Ask: *What prediction follows from this theory? What observation would contradict it? What result would require revision?*

This question is especially important for new foundational proposals. A theory should not be protected by changing its meaning whenever evidence disagrees. Reality must be allowed to answer.

9. What Lies Beneath the Present Explanation?

Physics often explains one layer by moving to another. Force is explained through fields. Fields may be described through geometry, quantum states, or symmetries. Temperature is explained through collective microscopic behaviour. Particles are described through quantum fields.

But every deeper description raises another question. What gives rise to the field? Why does the symmetry exist? Why does spacetime have its structure? How do stable matter forms emerge?

The purpose of asking what lies beneath is not to demand an infinite regress. It is to keep the present explanation from becoming prematurely final. Every scientific concept should remain open to deeper structure if evidence requires it.

10. What Is Reality Doing Before We Name It?

This may be the most important question in the book. Before we call something force, what is happening? Before we call something temperature, what physical state exists? Before we call something pressure, what collective interaction is occurring?

Before we speak of charge, field, particle, or entropy, what recurring manifestation has been observed? Reality does not wait for our vocabulary. It changes, interacts, stabilizes, propagates, transforms, and produces consequences. Science then develops terms to organize those consequences.

The vocabulary is necessary. But the physical event comes first. A science lover should therefore learn to look through the name rather than stop at it. The deepest question is not always: *What quantity should I calculate?* Sometimes it is: *What is reality doing that this quantity allows me to describe?*

A Final Question

These ten questions lead toward one larger question: *Are our present theories the final architecture of reality, or the most successful manifestation maps we have developed so far?*

We do not yet know. That uncertainty should not reduce our trust in science. It should deepen our participation in it. Science does not require us to pretend that every question has been settled. It requires us to observe carefully, measure honestly, reason clearly, mark our assumptions, test our ideas, and remain willing to revise them.

The true science lover does not reject answers. Nor does the science lover worship them. An answer is respected for what it explains. A theory is trusted where it works. A quantity is used where it measures reliably. And the question beneath them remains alive.

Key Reflections from the Book

Physics is not only a set of formulas—it is a way of seeing reality more deeply. These reflections remind us that beneath every equation lies a phenomenon, and beneath every phenomenon lies a question still alive.

1. **Reality existed before physics described it.** Objects moved, stars formed, heat transferred, and matter interacted long before human beings developed scientific quantities or equations.
2. **Physics begins in experience before it becomes mathematics.** Push, resistance, warmth, motion, sound, light, and weight were encountered through life before they were converted into definitions and units.
3. **Nature does not know scientific vocabulary.** A stone does not know its mass. A planet does not know its orbit. Matter does not read equations before responding to physical conditions.
4. **Physical laws do not command nature.** They are disciplined descriptions of regularities found within nature. The pattern makes the law possible; the law does not produce the pattern.
5. **The phenomenon is discovered, but the vocabulary is constructed.** Motion, interaction, and change are objective. The names, symbols, units, and conceptual categories used to describe them belong to science.
6. **A quantity is not necessarily a substance.** Force, temperature, pressure, charge, and entropy may be real as

measurable relations without existing as separate materials or independent agents.

7. **Measurement is a bridge between manifestation and knowledge.** An instrument responds physically. A scale and theory convert that response into a meaningful quantity.
8. **Every measurement selects.** It reveals one aspect of an event while leaving many others outside the immediate description.
9. **A number can be precise without being complete.** Accuracy tells us that a relation has been captured successfully. It does not guarantee that the entire physical process has been explained.
10. **The equation is not the event.** An equation represents relations among quantities. It does not contain the material, history, experience, or full ontology of the system.
11. **Prediction and explanation are not identical.** A theory may predict outcomes reliably while leaving open the deeper question of what produces them.
12. **Scientific theories are maps, not territories.** Their strength lies in preserving relevant structure, not in reproducing every feature of reality.
13. **A theory may be true within its domain without being final.** Newtonian mechanics remains powerful even though relativity provides a broader description. Higher-level truth can survive deeper explanation.
14. **Scientific progress often changes the meaning of familiar terms.** Force, matter, particle, field, vacuum, space, and time have all changed as physics moved into deeper layers.
15. **Force may describe response rather than an independent cause.** What physics calls force can arise through fields, gradients, constraints, deformation, momentum exchange, or geometric structure.
16. **Objects do not consciously "feel" forces.** They accelerate,

deform, exchange energy, redistribute momentum, and respond according to their organization and surroundings.

17. **Gravity shows that force may not be fundamental.** Newtonian physics describes gravitational attraction as force. General relativity describes free motion through curved spacetime.
18. **Contact forces are also deeper interactions.** Solidity, support, tension, friction, and elasticity arise from organized material and electromagnetic responses at microscopic scales.
19. **Fields deepened physics but did not end the inquiry.** They replaced mysterious action at a distance with distributed physical structure, yet the question of what fields fundamentally are remains open.
20. **The vacuum is not necessarily simple emptiness.** Removing ordinary matter does not remove fields, geometry, or every physical possibility.
21. **Matter may be stable form rather than static substance.** Persistence does not require the absence of change. A structure can remain identifiable while continuous dynamics occur within it.
22. **Stability is organized continuity.** A physical form survives when its internal relations maintain coherence through changing conditions.
23. **Patterns precede quantities.** Recurrence, symmetry, cycles, interference, and organization are noticed before they are converted into measurements.
24. **Manifestation is the observable face of deeper dynamics.** Physics often knows an underlying process through its consequences: motion, radiation, deformation, detector signals, or changing states.
25. **Consequence is continuous.** Every physical change modifies the conditions from which later changes arise. Cause and

effect are often parts of an ongoing relational process.

26. **Organisation matters as much as components.** The same materials arranged differently can produce entirely different physical capacities and behaviours.
27. **Emergent properties are real without being fundamental.** Temperature, pressure, elasticity, and other system-level properties depend on collective structure rather than isolated components.
28. **Physics is more than the study of how much.** It also asks how something becomes measurable, how structures emerge, why patterns persist, and how one descriptive layer arises from another.
29. **Mathematics reveals structure but does not interpret itself.** Symbols acquire physical meaning through definitions, measurements, assumptions, and experimental correspondence.
30. **Mathematical possibility is not physical confirmation.** A consistent solution may describe a real phenomenon, an approximation, an excluded state, or only a formal possibility.
31. **Beauty can guide physics, but it cannot replace evidence.** Elegance invites investigation. Nature decides whether the idea survives.
32. **Scientific language must mark levels of certainty.** Observed, measured, inferred, modelled, interpreted, and speculated are not interchangeable terms.
33. **Speculation is not the enemy of science.** Many important discoveries began with imagination. The problem begins when speculation is presented as established fact.
34. **Philosophy does not weaken physics.** It helps physics examine its assumptions, meanings, interpretations, and boundaries.
35. **The unmeasured is not automatically unreal.** It may be

irrelevant to the current problem, inaccessible to present instruments, or not yet formulated as a measurable variable.

36. **A deeper explanation should preserve what already works.** A new framework earns scientific value by recovering reliable results, clarifying unresolved questions, and producing testable consequences.
37. **A new name is not a new explanation.** Replacing force with field, field with information, or information with substrate does not solve the problem unless the physical mechanism becomes clearer.
38. **Structured Vacuum Energy is a proposed foundation, not a completed conclusion.** It offers the possibility that matter, fields, and physical manifestations arise from one continuous structured reality.
39. **SVE must not become another reified word.** Its value depends on definitions, mathematical development, correspondence with known physics, distinctive predictions, and openness to failure.
40. **Physical quantities may be observer vocabulary for deeper processes.** They translate recurring manifestations into forms that can be compared, communicated, and calculated.
41. **Observer vocabulary does not make nature subjective.** Reality constrains every reliable measurement, even though the units, symbols, and descriptive categories are human constructions.
42. **The universe is one; scientific descriptions are many.** Nature does not divide itself into mechanics, thermodynamics, relativity, quantum theory, and cosmology. Those divisions help human inquiry.
43. **The deepest question is often hidden beneath the correct answer.** A numerical result may be right while the physical meaning remains open.

44. **Science should remain confident without becoming final.** Established theories deserve trust where they work, but no successful framework should be protected from deeper questioning.
45. **Wonder and rigour belong together.** Wonder opens the question. Mathematics disciplines the proposal. Experiment tests the consequence. Reality gives the response.
46. **The future of physics may depend on connecting levels.** The next advance may explain how substrate becomes field, field becomes matter, microscopic dynamics becomes macroscopic form, and manifestation becomes measurement.
47. **A correct description is not the same as complete possession of reality.** Scientific knowledge can be objective, powerful, and incomplete at the same time.
48. **Reality is always larger than the vocabulary used to describe it.** The purpose of physics is not merely to name the universe, but to understand what the names allow us to see—and what they may still conceal.

Final Reflection

The central invitation of this book is not to distrust physics.

It is to look more deeply through it.

Use the quantities. Learn the equations. Respect the theories. Understand their domains.

But continue asking what physical process makes the quantity possible, what structure lies beneath the equation, and what part of the explanation remains interpretive.

Physics becomes smaller when it is reduced to numerical problems. It becomes vague when it abandons measurement. Its true depth lies between these extremes.

It is the disciplined study of how reality changes, organizes, interacts, manifests, and becomes measurable.

The numerical answer may complete the problem. The deeper question continues the science.

A Short Note on UED and SVE

Universal Energy Dynamics, or **UED**, is a developing conceptual framework that attempts to understand physical reality beneath the familiar vocabulary of objects, forces, fields, and measurable quantities.

Its starting point is simple:

Reality may be continuous before it appears as separate things.

What we call matter, fields, motion, force, and physical properties may not be independent ingredients placed inside an empty universe. They may be different manifestations of one deeper structured foundation.

Within UED, this proposed foundation is called **Structured Vacuum Energy (SVE)**.

Understanding the Term

SVE is not intended to mean empty space containing a hidden mechanical substance. It is not simply a modern renaming of the classical ether. Nor should it be imagined as an invisible fluid filling a pre-existing container called space.

Instead, SVE refers to the possibility that the physical vacuum possesses an underlying structured and dynamic character from which matter, fields, interactions, and perhaps spacetime relations themselves may emerge.

- The word **structured** is essential. A completely featureless foundation could not easily account for stable differences, organized forms, gradients, waves, particles, fields, or persistent physical behaviour.

- The word **energy** is used cautiously. In established physics, energy is a measurable quantity associated with the state and transformation of systems. In UED, the term points more broadly toward the dynamic capacity of the foundation to change, organize, propagate disturbances, and form stable configurations.

This broader use must eventually be connected precisely with standard definitions. Without such formal development, the word *energy* risks becoming too vague to carry scientific meaning.

Four Core Propositions of UED

1. **Matter as stable form.** Matter may be understood as organized continuity rather than absolute stillness. A stone, atom, or particle is not frozen substance but a persistent configuration of deeper activity.
2. **Fields as distributed conditions.** Matter and fields may be different modes of one continuous substrate—localized versus extended manifestations of the same foundation.
3. **Force as manifestation variable.** Force may represent measurable response rather than fundamental cause. It remains indispensable in mechanics, but its deepest origin may lie in gradients, geometry, or substrate dynamics.
4. **Quantities as observer vocabulary.** Mass, charge, temperature, pressure, momentum, and entropy are objective and measurable, yet they are also conceptual tools developed by observers to describe recurring manifestations.

A Proposed Hierarchy

- Structured reality

- Dynamic difference
- Interaction and response
- Stable or changing manifestation
- Observation
- Measurement
- Physical quantity
- Equation
- Theory
- Interpretation

This hierarchy clarifies where physical quantities and theories stand in relation to reality. It does not place UED above established physics. Current physics remains the necessary foundation for any serious development of UED.

Scientific Standards

At present, UED should be regarded as a **research framework and foundational proposal**, not as a completed replacement for modern physics. Its development requires:

- Clear definitions
- Mathematical formulation
- Correspondence with established observations
- Distinctive predictions
- Experiments capable of supporting or challenging those predictions

Without these, UED remains philosophical rather than physical.

The Value of UED

The framework's strength lies in the questions it organizes:

- Can matter be understood as stable energy structure?
- Can fields be modes of one universal foundation?
- Can force emerge from gradients and responses rather than exist as a primary agent?
- Can physical quantities be derived from deeper dynamics?
- Can continuity reinterpret the apparent separation between object and environment?
- Can a substrate model produce testable differences from existing field descriptions?

These questions deserve investigation. They do not yet possess final answers.

A Responsible Doorway

This book has used UED only as a doorway. The main argument does not depend on accepting SVE. Even without UED, the distinction between reality and description remains important:

- Physical quantities remain scientific vocabulary.
- Equations remain representations.
- Fields remain conceptually open.
- The vacuum remains more complicated than ordinary emptiness.

UED enters after these distinctions have been established. It offers one possible answer to the question:

What kind of physical reality might allow matter, fields, interaction, and measurable quantities to emerge together?

The future value of UED will depend on whether that possible answer can become testable physics.

Closing Note

I have taught physics for many years, and during that time I have seen students solve thousands of numerical problems.

They learn formulas. They substitute values. They draw diagrams. They calculate force, energy, velocity, pressure, temperature, and momentum.

Some become very fast at solving problems. Some score excellent marks. Some move ahead into engineering, medicine, research, and technology.

But every now and then, a student pauses after obtaining the correct answer and asks something more basic:

"Sir, what is actually happening here?"

That question has always interested me more than the calculation.

Not because the calculation is unimportant—it is essential. Physics without measurement and mathematics would lose its precision. But the numerical answer often closes the textbook problem before it opens the physical problem.

A student may calculate gravitational force correctly and still not know what gravity means. A student may use the equation for pressure without seeing the collective interaction hidden beneath the quantity. A student may learn that temperature measures a thermal state without asking how a system of moving particles becomes one measurable condition. A student may repeat that particles are excitations of fields without ever asking what a field is.

This book grew from such questions.

It did not begin as an attempt to reject physics. I have spent much of my life teaching it, using its equations, trusting its predictions, and showing students how deeply it shapes technology and modern life.

My concern is different.

I feel that physics is often introduced only after its wonder has been converted into vocabulary. Students meet definitions before dilemmas. They meet formulas before experiences. They meet laws before learning why those laws became necessary. They are taught what to calculate, but not always what the calculation represents.

Over time, the symbols become more familiar than the reality. Force becomes *(F)*. Energy becomes *(E)*. Mass becomes *(m)*. The student learns to manipulate the letters and may gradually forget that each symbol emerged from a long human attempt to understand recurring physical consequences.

I wanted to return to the event before the symbol.

What happens before we call something force? What exists before we assign temperature? What are we measuring when we speak of pressure, charge, mass, or entropy? Does nature contain these quantities exactly as our textbooks present them, or are they the vocabulary through which human intelligence organizes observable behaviour?

These questions may sound philosophical, but physics has never been separate from philosophy. Every major transformation in physics changed not only equations but also meaning.

Newton changed the relation between force and motion. Faraday changed the status of the space around objects. Maxwell revealed unity where separate phenomena had been seen. Einstein changed the meanings of space, time, motion, and gravity. Quantum physics changed the meaning of particle, state, measurement, and probability.

Science moves forward when someone asks whether the familiar description is the final one.

The purpose of this book has been to preserve that permission to ask.

I do not expect every reader to agree with every interpretation offered here. Nor do I ask the reader to accept Universal Energy Dynamics or Structured Vacuum Energy merely because they provide a broader picture.

A new framework should not be protected from criticism. It should be clarified through criticism. It should be tested. It should be revised. It should be rejected where it fails and strengthened where consequences support it.

I have tried to mark that boundary honestly.

There is a difference between observation and inference. There is a difference between an equation and its interpretation. There is a difference between mathematical possibility and physical confirmation. There is a difference between a meaningful conceptual proposal and an established scientific theory.

These differences should not discourage imagination. They should discipline it.

My intention in introducing SVE is not to replace the authority of one vocabulary with another. The term is useful only if it helps us ask more precise questions about matter, fields, vacuum, stability, and physical manifestation. If it becomes only another word that appears to explain everything, then it has failed.

The real work begins when the word is forced to produce structure, mathematics, predictions, and experiments.

This book is therefore not a declaration that the deepest reality has been found. It is a statement that the search should remain open.

Physics has become extraordinarily powerful, but its power should not make it intellectually closed. Theories can predict accurately while deeper meanings remain unresolved. Quantities can be real and useful without being fundamental. Equations can reveal structure without becoming the reality they describe. A field can be central to a theory while its ultimate physical nature remains open.

There is no insult to science in saying this. Science is strongest when it knows the boundary between what it has measured and what it has interpreted.

As a teacher, I have never wanted students merely to remember what I say. I want them to learn how to look.

A teacher gives a formula, but should also show the situation from which it arose. A teacher explains a law, but should also reveal its conditions and limits. A teacher provides an answer, but should not destroy the student's ability to ask a better question.

The future of science will not be created only by those who solve existing problems quickly. It will also be created by those who notice that the accepted problem has been framed too narrowly.

Perhaps one reader of this book will return to force and see response. Perhaps another will return to matter and see stable organization. Another may look at the vacuum and no longer imagine simple emptiness. Another may develop a better mathematical framework than the one proposed through UED. Another may identify a weakness that requires the entire idea to be revised.

All these would be useful consequences. Knowledge should not become the private property of its author. Once an idea enters the shared field of inquiry, it must be allowed to meet stronger minds, better instruments, new evidence, and different interpretations.

I have written this book not because I believe every question has been answered, but because I believe some important questions are no longer being asked often enough.

What is force before it becomes a quantity? What is matter before it appears solid? What is a field beyond its mathematical definition? What is the vacuum when ordinary matter is absent? Why does reality form stable patterns? Why can human mathematics describe those patterns? How does a physical process become a measurable manifestation?

Perhaps these questions will eventually lead toward SVE. Perhaps they will lead somewhere else.

The destination should not be decided before the journey.

For the student, teacher, engineer, scientist, or science lover reading these final pages, I leave only one suggestion:

Do not stop using equations. Do not stop trusting measurement. Do not stop learning established physics.

But do not stop at the equation either.

Look at the event. Look at the assumptions. Look at the scale. Look at what has been measured and what has merely been named. Look at what the theory explains and what it leaves open.

Then ask, once more:

What is reality actually doing here?

That question has accompanied me throughout my years of teaching physics. It still has not become old.

And as long as that question remains alive, physics will remain more than we know.

— Sandeep Chavan

Glossary of Core Terms

Acceleration

The rate at which velocity changes with time. In classical mechanics, acceleration is related to the net force acting on a body and its mass. In this book, it is also treated as an observable response that may arise from deeper fields, gradients, constraints, or geometry.

Action at a Distance

The idea that one body can influence another across space without an obvious intervening mechanism. Newtonian gravity raised this question because masses appeared to attract one another through empty space. Field theory later provided a more local description of interaction.

Approximation

A simplified description that remains sufficiently accurate under specified conditions. Newtonian mechanics, for example, is an excellent approximation for many ordinary motions even though relativity provides a broader framework.

Boundary Condition

A constraint imposed on a physical system at its edges, limits, or initial state. Boundary conditions help determine which behaviours or solutions are physically possible within a model.

Causation

The relation through which one condition, interaction, or event contributes to another. This book broadens causation beyond the simple picture of one object pushing another and considers structural, relational, and continuous consequence-based causation.

Charge

A measurable physical property associated with electromagnetic interaction. Positive and negative charge describe opposite modes of electromagnetic response. The quantity is experimentally well established, while its deepest physical origin remains open to interpretation.

Classical Physics

The body of physical theories developed before relativity and quantum mechanics, especially Newtonian mechanics, classical electromagnetism, and classical thermodynamics. Classical physics remains highly effective within its proper domains.

Coherence

The maintenance of an organized relation among parts of a system. Coherence may refer to phase relations, structural stability, or the ability of a pattern to persist through changing conditions. In this book, the term is used broadly and should not always be confused with its narrower technical meanings in quantum physics.

Configuration

The arrangement or state of the components, fields, or variables within a physical system. Different configurations can produce different energies, responses, and observable manifestations.

Consequence

A physical change produced through prior conditions, relations, or interactions. A consequence can become the condition for later events, creating a continuous chain of physical evolution.

Constraint

A condition that limits the possible motion, form, or evolution of a system. A floor constrains a falling body, a container constrains a gas, and material structure constrains deformation.

Continuity

The idea that physical reality may possess connected structure rather than consist only of completely separate objects. Within UED, continuity is a proposed feature of the deeper substrate from which matter and fields arise.

Description

A conceptual, verbal, mathematical, or graphical representation of physical behaviour. A description may be accurate and useful without being identical to the reality it represents.

Determinism

The view that a complete physical state and governing laws uniquely determine all later states. Classical mechanics encouraged strong deterministic interpretations, while quantum theory complicated this picture through probabilistic predictions.

Dynamics

The study or process of change within a physical system. Dynamics includes motion, interaction, energy redistribution, field variation, deformation, and transformation.

Effective Theory

A theory that accurately describes phenomena within a limited range of scales or conditions without necessarily being the deepest possible explanation. Classical mechanics and thermodynamics are important examples.

Emergence

The appearance of stable properties or behaviours at one level that arise from collective organization at a deeper level. Temperature and pressure are emergent because they describe system-level conditions rather than isolated particles.

Energy

A measurable physical quantity associated with the state, transformation, and capacity for change of a system. Energy should not automatically be imagined as a material substance. Within UED, the word also points toward the dynamic capacity of the proposed underlying structure.

Energy Gradient

A change in energy or potential across position, configuration, or state. Systems often respond to such differences, producing motion, transfer, redistribution, or other physical changes.

Entropy

A thermodynamic and statistical quantity associated with energy distribution, accessible configurations, and the direction of physical processes. It should not be reduced casually to "disorder," although that metaphor can sometimes be useful.

Epistemology

The study of knowledge: how claims are formed, justified, tested, and distinguished from belief or interpretation. In physics, epistemology asks what has been observed, inferred, modelled, or speculated.

Evidence

Observed or measured information used to support, challenge, or refine a scientific claim. Evidence does not always determine a single interpretation, but it constrains which interpretations remain scientifically credible.

Field

A physical or mathematical structure that assigns values or conditions throughout space and time. Fields allow interactions to be described locally. Modern physics treats fields as central, though their ultimate ontology remains open.

Force

A physical quantity used to describe changes in momentum, acceleration, deformation, or mechanical interaction. Force remains fundamental to practical mechanics, but this book questions whether it is ontologically fundamental.

Free Fall

Motion under gravitational influence without supporting contact forces. In general relativity, an ideal freely falling body follows a natural path through spacetime and experiences no proper acceleration.

Fundamentality

The status of something treated as irreducible within a theory. A concept may be fundamental in one framework and emergent in a deeper framework. Force is fundamental in parts of classical mechanics but not in the same way in general relativity.

Geometry

The study of spatial and relational structure. In general relativity, geometry is physically dynamic: matter and energy are associated with spacetime curvature, which shapes motion.

Gradient

A mathematical expression of how a quantity changes across space or configuration. Temperature, pressure, concentration, and potential gradients are associated with directional physical responses.

Information

A description of distinguishable states, differences, or configurations. Information must be physically embodied in some state or structure and should not automatically be treated as a substance existing independently of physical reality.

Interaction

A relation in which physical systems influence one another or exchange energy, momentum, or other measurable properties. Interactions may be described through forces, fields, couplings, collisions, or changing configurations.

Interpretation

An account of what a scientific model or mathematical structure is believed to mean physically. Several interpretations can sometimes share the same experimentally successful formalism.

Law of Nature

A concise description of a recurring physical regularity. Physical laws do not command nature in the manner of human laws; they summarize reliable patterns found through observation and experiment.

Manifestation

The observable or measurable expression of an underlying physical condition or process. Motion, radiation, deformation, temperature, pressure, and detector signals are examples of physical manifestations.

Manifestation Science

A proposed broader view of physics that studies not only measurable quantities, but also how deeper structures and dynamics become observable and measurable. It supplements rather than replaces quantitative science.

Mass

A measurable property associated with inertia, energy, and gravitational behaviour. In classical mechanics, mass expresses resistance to acceleration. In deeper theories, its meaning becomes connected with fields, internal energy, and physical organization.

Matter

The stable physical forms that make up observable objects and systems. In conventional physics, matter is described through particles and fields. In UED, matter is approached as a persistent structured form of SVE.

Measurement

The comparison of a physical manifestation with an agreed standard through an instrument or procedure. Measurement creates a disciplined relation between a physical system and scientific vocabulary.

Model

A simplified conceptual, mathematical, or computational representation of a physical system. Models select relevant features and omit others so that behaviour can be analysed or predicted.

Momentum

A physical quantity associated with motion, commonly expressed in classical mechanics as mass multiplied by velocity. Momentum is conserved in isolated systems and can be transferred through interaction.

Observer

A reference frame, measuring system, apparatus, or conscious interpreter involved in the registration and description of physical events. In this book, the term does not imply that conscious attention creates every physical event.

Observer Vocabulary

The system of quantities, symbols, units, and concepts developed by observers to represent recurring physical manifestations. The vocabulary is humanly constructed but objectively constrained by nature.

Observation

The registration of a physical event through perception or instrumentation. Observation should be distinguished from the theoretical interpretation later assigned to the recorded event.

Ontology

The study of what fundamentally exists. In physics, ontology asks whether particles, fields, spacetime, information, or a deeper substrate should be treated as physically real and in what sense.

Particle

A localized quantum or physical entity associated with characteristic properties such as mass, charge, and spin. In quantum field theory, particles are commonly described as excitations of underlying fields rather than miniature classical objects.

Pattern

A recurring or organized relation in space, time, behaviour, or data. Patterns often become scientifically visible before quantities are developed to measure them precisely.

Physical Quantity

A defined measurable property expressed through a numerical value and unit. Physical quantities form the operational vocabulary through which science compares and mathematically relates manifestations.

Potential

A quantity describing the capacity of a system to produce change according to position or configuration. Changes in potential are often connected with force-like responses or energy transfer.

Prediction

A statement about an observable outcome derived from a theory or model. Successful prediction strengthens confidence that a theory captures real physical structure, but it does not necessarily establish one final interpretation.

Pressure

A measurable quantity describing force distributed over area. In gases and fluids, pressure often emerges from collective microscopic interactions and momentum transfer.

Probability

A measure of the likelihood of possible outcomes. In quantum physics, probabilities are central to the formalism, although their deeper interpretation remains debated.

Quantum Field

A field described through quantum principles whose excitations are associated with particles. Quantum fields differ fundamentally from simple classical material media.

Reality

The physical existence and processes that precede human description. Reality constrains observation, measurement, theory, and interpretation but is not assumed to be exhausted by any one vocabulary.

Reification

The mistake of treating an abstraction, quantity, or relation as though it were a separate concrete object. Speaking as though force, entropy, or information acts independently can produce reification.

Relativity

The physical theories developed by Albert Einstein that transformed the concepts of space, time, motion, energy, and gravity. General relativity describes gravity through spacetime geometry rather than as an ordinary force.

Response

The change in motion, structure, state, or behaviour of a physical system under particular conditions. This book often uses response as a deeper and less agent-like alternative to saying that a force "acts."

Scale

The range of size, energy, time, or complexity at which a system is studied. Different quantities and theories become useful at different scales.

Spacetime

The unified geometrical structure of space and time used in relativity. Spacetime is dynamic in general relativity and can be curved by matter and energy.

Speculation

A proposed explanation or possibility not yet fully supported by evidence. Speculation can guide research when clearly identified, but it should not be presented as confirmed fact.

Stability

The ability of a system or form to persist, return toward a state, or resist disruptive change. Stability does not require complete stillness; it may arise through continuing organized dynamics.

Structured Vacuum Energy (SVE)

The proposed underlying physical foundation within UED. SVE is conceived as continuous, dynamic, and structured rather than as empty nothingness or a simple mechanical medium. Matter and fields are proposed to arise as organized manifestations of this foundation.

Substrate

An underlying physical basis from which observable forms, fields, or relations may arise. A scientifically meaningful substrate must eventually be defined mathematically and connected with testable consequences.

Symmetry

A property of a physical or mathematical system that remains unchanged under particular transformations. Symmetries play a central role in modern physics and are closely connected with conservation laws.

Temperature

A measurable thermodynamic quantity describing the thermal state of a system. It emerges from collective energetic and statistical behaviour rather than existing as a simple substance inside matter.

Theory

An organized explanatory framework connecting concepts, mathematical relations, observations, and predictions. A scientific theory is stronger than a guess, but it may still possess limits and unresolved interpretations.

Universal Energy Dynamics (UED)

A developing foundational framework proposing that matter, fields, interactions, and physical manifestations arise through the structured dynamics of a universal substrate. UED is presented in this book as a research direction rather than completed or established physics.

Vacuum

A region with little or no ordinary matter. In modern physics, the vacuum is not necessarily equivalent to absolute nothingness; fields and physical structure may remain present.

Velocity

The rate of change of position with direction. Velocity differs from speed because it includes directional information.

Wave

A propagating disturbance or recurring pattern through which energy, momentum, or information may be transmitted. The physical meaning of a wave depends on the framework: water waves, electromagnetic waves, and quantum wavefunctions are not identical kinds of entities.

Final Note on Usage

Many terms in physics carry more than one meaning.

A word may have an everyday meaning, a precise operational definition, a mathematical role, and a broader philosophical interpretation. Confusion begins when these layers are combined without explanation.

The purpose of this glossary is not to provide final definitions for all of physics. It is to help the reader recognize which layer of meaning is active in the book.

A term can be scientifically useful without being ontologically final.

A quantity can be measurable without being a substance.

A theory can be accurate without exhausting reality.

And a new vocabulary becomes meaningful only when it leads back to observable consequence.

Further Reading

From the Chavanian Ecosystem and Beyond

This book stands at the intersection of physics, philosophy, scientific language, and foundational inquiry. It does not attempt to replace standard textbooks, technical monographs, or established physical theories. Its purpose is different: to help the reader examine what physical quantities mean, how theories represent reality, where interpretation enters science, and why deeper questions remain open even after a successful calculation has been completed.

Readers who wish to continue this inquiry may follow two complementary paths.

- The first path moves further into the **Chavanian ecosystem**, where physics is connected with energy dynamics, consequence, consciousness, psychology, society, leadership, and human behaviour.
- The second path moves outward into **established works of physics and philosophy of science** that have shaped the modern understanding of matter, fields, space, time, measurement, emergence, and scientific explanation.

The two paths should not be treated as competitors. The Chavanian works offer a developing integrative framework. The established works provide the historical, scientific, and philosophical foundations against which every new proposal must be examined.

Reading Further Within the Chavanian Ecosystem

Universal Energy Dynamics

Readers seeking the broader physical foundation behind this book should begin with the central works on **Universal Energy Dynamics (UED)**. These works develop the proposal that reality is grounded in a continuous structured energetic foundation and that matter, fields, forces, and measurable physical quantities may arise as manifestations of deeper dynamics.

They explore concepts such as Structured Vacuum Energy, ripple behaviour, alignment, matter formation, field response, and the relation between energy structure and physical consequence. This is the most direct continuation for readers interested in the SVE doorway introduced in the later chapters of the present book.

Recommended sequence:

1. Introductory works explaining the conceptual foundations of UED
2. Papers or monographs on Structured Vacuum Energy
3. Works applying UED to force, fields, quantum behaviour, cosmology, and matter formation
4. Later technical works proposing mathematical or experimental models

Readers should approach these works as part of a developing research programme rather than as a closed doctrine. Their value lies in the questions they organize and the possible connections they attempt to build across physical domains.

The Law of Consequences

The idea of consequence appears repeatedly in this book. Physical reality does not unfold as isolated events; every change modifies the conditions from which later changes arise.

The Law of Consequences develops this view more broadly. It examines consequence not only as an outcome, but as the continuing architecture through which physical, biological, psychological, and social systems evolve.

For readers of the present book, this work extends the transition from simple linear causation toward relational and continuous consequence. It is especially relevant to the chapters on manifestation, energy dynamics, causation, and the future of physics.

Thermodynamics of Thoughts

This work carries thermodynamic language into the domain of cognition while remaining centred on pattern, energy cost, unresolved delta, oscillation, expression, and resolution.

It does not claim that thoughts are identical to heat or mechanical systems. Instead, it investigates whether thermodynamic principles can illuminate how cognitive patterns arise, persist, accumulate, and seek resolution.

Readers interested in how physical concepts migrate into psychology and philosophy may find this work useful. It also demonstrates the need for caution when transferring scientific vocabulary across domains. An analogy becomes valuable only when its limits remain visible.

From Synapse to Self

This work extends foundational inquiry from physics into neuroscience, consciousness, and the emergence of self. It examines the distinction between brain activity, pattern generation, meaning, awareness, and the self as an organizing or resolving anchor.

The book asks whether neural processes alone fully explain experience or whether a broader field-based account is required. Readers interested in the observer, measurement, awareness, and the relation between physical process and experienced meaning may continue here.

Its central question is similar to one raised throughout the present book: *What lies between measurable activity and the reality we say the measurement represents?*

The Mind That Won't Stop

This is a more practical and human-facing exploration of thought, attention, control, and mental pressure. Where *Physics: More Than You Know* asks whether quantities are mistaken for reality, this work asks whether thoughts are mistaken for the self.

Both books examine the danger of confusing representation with source. It is especially relevant for readers interested in how language, interpretation, and internal modelling shape human experience.

What Is Scattering You?

This shorter work examines attention, distraction, unfinished thought, comparison, digital noise, and the loss of inner coherence. It is useful for younger readers and for those who prefer a practical entry into the broader Chavanian framework.

Its connection with the present book lies in the concept of organization. A physical system behaves differently according to structure and alignment. Human attention also changes according to how internal and external demands are organized.

Know Yourself Before Life Misreads You

This work explores identity, self-understanding, social interpretation, and the difference between the person and the labels attached to the person.

It parallels one of the main claims of this book: *the name is not the reality*. Just as force, mass, and field can become reified through scientific language, human roles and social identities can be mistaken for the deeper self.

Readers interested in the wider philosophical consequences of representation may find this connection valuable.

Physics of Love and Sex

This work applies a field-and-consequence perspective to attraction, attachment, intimacy, embodiment, and relational coherence. It is not a physics textbook, nor does it reduce love to a mechanical quantity.

Its purpose is to explore how human relationships can be understood through interaction, embodiment, energetic consequence, attachment, and shared fields. Readers should approach it as an interdisciplinary application of the wider Chavanian framework.

Chavanian Leadership

This work examines leadership not as a permanent personal trait but as a temporary field function. It proposes that leadership emerges when one person becomes capable of sensing, stabilizing, or resolving a shared delta within a group.

The connection with the present book lies in the movement from object to relation. Just as force may be understood as response rather than an independent entity, leadership may be understood as an emergent function rather than a substance possessed by an individual.

Constitutional Geometry

This work applies structural and field-based reasoning to constitutions, institutions, power, rights, authority, and collective organization.

It asks how a society converts shared principles into operational structures and why systems fail when their constitutional geometry becomes misaligned with lived reality. Readers interested in how physical ideas such as structure, boundary, stability, and consequence can inform institutional thought may continue here.

The Game of R's

This work develops a sequence of Resistance, Rebellion, Reform, Revolution, and Resolution. It explores how systems respond when pressure accumulates and existing structures can no longer absorb change.

Its relevance to the present book lies in dynamic transition. Physical systems cross thresholds. Social systems do the same. The language differs, but the underlying concern remains: *how does a system move from one structured condition into another?*

Reading Beyond the Chavanian Ecosystem

The following works are recommended not because they necessarily support UED, but because they deepen the reader's understanding of established physics, scientific interpretation, and the history of foundational questions. A serious reader should engage with ideas that confirm, challenge, and complicate the framework presented in this book.

Classical Mechanics and the Scientific Revolution

Isaac Newton — *The Principia: Mathematical Principles of Natural Philosophy* Newton's work remains foundational to classical mechanics and gravitational theory. Readers need not master every mathematical detail to appreciate the scale of the conceptual achievement: terrestrial and celestial motion were brought into one quantitative framework. It is also useful to observe what Newton's mathematics explained and what remained open regarding the nature of gravitational action.

Galileo Galilei — ***Two New Sciences*** Galileo's work reveals the conceptual struggle required to separate motion from ordinary assumptions about force, rest, resistance, and falling bodies. It is valuable for understanding how experiment, idealization, and mathematical reasoning began transforming physical inquiry.

Ernst Mach — ***The Science of Mechanics*** Mach examined the historical and conceptual foundations of mechanics with unusual critical attention. His work is useful for readers interested in how scientific quantities arise from experience and how apparently obvious mechanical concepts can conceal philosophical assumptions.

Fields, Electricity, and Light

Michael Faraday — ***Experimental Researches in Electricity*** Faraday's work demonstrates how experimental insight can precede mature mathematical formalism. His thinking about lines of force helped transform the space around objects from passive emptiness into physically significant structure.

James Clerk Maxwell — ***A Treatise on Electricity and Magnetism*** Maxwell mathematically unified electricity, magnetism, and light. His work is central to understanding the rise of the field concept and the extraordinary capacity of mathematics to reveal connections not obvious from isolated observations.

Albert Einstein — **"On the Electrodynamics of Moving Bodies"** Einstein's 1905 paper on special relativity transformed the relation among space, time, motion, and electromagnetism. It remains an important example of how a conceptual reorganization can become more powerful than adding corrections to an older framework.

Relativity, Gravity, and Spacetime

Albert Einstein — ***Relativity: The Special and the General Theory*** This is one of the most accessible direct introductions to Einstein's conceptual framework. It helps readers see why gravity could no longer remain merely a force acting within fixed space and time.

Hermann Minkowski — "Space and Time" Minkowski's formulation of spacetime provided a geometrical structure for special relativity. It is useful for understanding how geometry became part of physical explanation rather than only a mathematical tool.

John Archibald Wheeler — ***A Journey into Gravity and Spacetime*** Wheeler presents the conceptual significance of general relativity in an accessible style. His work helps readers appreciate gravity as geometry while also seeing the questions that remain beyond the geometric description.

Quantum Theory and Its Interpretation

Niels Bohr — ***Atomic Physics and Human Knowledge*** Bohr's essays examine quantum theory, measurement, complementarity, and the limits of ordinary language. They are important for readers interested in how observation and description enter physical theory.

Werner Heisenberg — ***Physics and Philosophy*** Heisenberg reflects on the conceptual transformation produced by quantum mechanics. The book is particularly relevant to the distinction between mathematical formalism, observation, and interpretation.

Erwin Schrödinger — ***What Is Life?*** **and** ***Science and Humanism*** Schrödinger's writings cross the boundaries among physics, biology, philosophy, and human meaning. They demonstrate how a major physicist can remain mathematically serious while asking questions broader than calculation.

John von Neumann — ***Mathematical Foundations of Quantum Mechanics*** This is a more technical work, but it is historically significant for the mathematical structure of quantum theory and the measurement problem.

David Bohm — *Quantum Theory* and *Wholeness and the Implicate Order* Bohm's works are useful for readers interested in alternatives to purely operational interpretations of quantum mechanics. His later philosophical writing emphasizes wholeness, process, and deeper order, though readers should distinguish carefully between formal physics and philosophical extension.

John S. Bell — *Speakable and Unspeakable in Quantum Mechanics* Bell's essays are essential for understanding the conceptual and experimental importance of quantum nonlocality and the limits of certain hidden-variable theories.

Quantum Fields and the Vacuum

Richard P. Feynman — *QED: The Strange Theory of Light and Matter* Feynman provides an accessible introduction to quantum electrodynamics while preserving the strangeness of the theory. The book demonstrates how highly accurate predictions can coexist with conceptual difficulty.

Steven Weinberg — *The Quantum Theory of Fields* This is a technical reference for readers seeking a rigorous account of modern quantum field theory. It should be approached after substantial mathematical preparation.

Anthony Zee — *Quantum Field Theory in a Nutshell* Zee offers a conceptually energetic introduction to quantum field theory and its mathematical architecture.

Peter W. Milonni — *The Quantum Vacuum* This work explores the physical and theoretical significance of vacuum states in quantum field theory. It is especially relevant to readers interested in how modern physics differs from the classical idea of empty space.

Thermodynamics, Entropy, and Statistical Physics

Sadi Carnot — ***Reflections on the Motive Power of Fire*** Carnot's work is foundational to the development of thermodynamics and the study of heat engines. It reveals how scientific understanding can advance through operational relations before deeper microscopic explanations are established.

Rudolf Clausius — ***Selected Writings on Thermodynamics*** Clausius played a central role in the development of entropy and the second law of thermodynamics. His work helps readers trace how a technical quantity gradually acquired broader scientific and philosophical meaning.

Ludwig Boltzmann — ***Lectures on Gas Theory*** Boltzmann's work connects thermodynamic behaviour with statistical and microscopic processes. It is central to understanding emergence, probability, entropy, and the relation between levels of description.

Ilya Prigogine — ***Order Out of Chaos*** Prigogine explores irreversible processes, self-organization, and structures arising far from equilibrium. The work is useful for thinking about how order, change, and complexity can coexist.

Mathematics and Physical Reality

Eugene P. Wigner — "The Unreasonable Effectiveness of Mathematics in the Natural Sciences" This influential essay examines the surprising success of mathematics in describing physical reality. It is directly relevant to the chapter arguing that mathematics is powerful without being identical to reality.

Roger Penrose — ***The Road to Reality*** Penrose provides an extensive journey through mathematics and modern physics. The book is demanding but valuable for readers who wish to understand how mathematical structures shape contemporary physical thought.

Max Tegmark — ***Our Mathematical Universe*** Tegmark presents the strong view that physical reality may itself be mathematical.

Readers of the present book may find it useful as a contrasting position, particularly when considering the difference between mathematical structure and physical ontology.

Philosophy and Method of Science

Karl Popper — *The Logic of Scientific Discovery* Popper emphasizes testability and falsifiability as central features of scientific theories. His work is especially relevant to the question: *what observation would show a theory to be wrong?*

Thomas S. Kuhn — *The Structure of Scientific Revolutions* Kuhn examines how scientific paradigms develop, stabilize, encounter anomalies, and undergo transformation. His work helps explain why scientific progress is not always a smooth accumulation of facts.

Imre Lakatos — *The Methodology of Scientific Research Programmes* Lakatos provides a more flexible account of how research frameworks develop and how they should be judged over time. This is particularly relevant to a developing framework such as UED.

Pierre Duhem — *The Aim and Structure of Physical Theory* Duhem explores the nature of physical theory, experiment, representation, and the limits of isolated testing. His work is valuable for understanding theories as structured systems rather than direct photographs of reality.

Bas van Fraassen — *The Scientific Image* Van Fraassen presents a powerful argument for scientific empiricism and cautions against assuming that every successful theoretical entity must be interpreted as literally real.

Nancy Cartwright — *How the Laws of Physics Lie* Cartwright examines how physical laws work through idealization and limited domains rather than functioning as universal descriptions in every concrete situation. The provocative title should not be mistaken for a rejection of science; the book is a study of how models gain usefulness.

Ian Hacking — *Representing and Intervening* Hacking explores the relationship among scientific representation, experimentation, and realism. The book is particularly relevant to the distinction between describing an entity and interacting with it experimentally.

Emergence, Complexity, and Organisation

Philip W. Anderson — "More Is Different" Anderson's essay argues that large-scale organization can produce genuinely new physical behaviour not captured simply by listing microscopic components. It is a foundational statement on emergence.

Geoffrey West — *Scale* West examines patterns of scaling across organisms, cities, and complex systems. The work demonstrates how quantitative regularities can emerge across apparently different domains.

Melanie Mitchell — *Complexity: A Guided Tour* Mitchell offers an accessible introduction to complexity, emergence, computation, adaptation, and system-level behaviour.

Stuart Kauffman — *At Home in the Universe* Kauffman explores self-organization and the emergence of order in complex systems. His work is useful for readers considering stability and organization as physically significant features.

Information and Physical Reality

Claude Shannon — "A Mathematical Theory of Communication" Shannon's work established a rigorous mathematical treatment of information. It is important to read because later uses of the word *information* often extend far beyond Shannon's original technical framework.

Rolf Landauer — *Selected Essays on the Physical Nature of Information* Landauer argued that information is physically embodied and that logical operations can carry thermodynamic consequences.

His work helps prevent information from being treated as an abstract substance detached from physical systems.

John Archibald Wheeler — Writings on "It from Bit" Wheeler's speculative proposal that physical reality may arise from informational distinctions has influenced later thinking about quantum theory and cosmology. It should be read as a provocative foundational idea rather than a completed physical theory.

Cosmology and Foundational Questions

Stephen Hawking and George F. R. Ellis — *The Large Scale Structure of Space-Time* This is a technical foundational work on relativity and cosmology.

Steven Weinberg — *The First Three Minutes* Weinberg presents early-universe cosmology in a clear and scientifically disciplined way.

Lee Smolin — *The Trouble with Physics* Smolin examines unresolved questions in fundamental physics, research culture, and the search for unification. Readers may find it useful when considering why scientific institutions favour some programmes over others.

Carlo Rovelli — *Reality Is Not What It Seems* Rovelli offers an accessible journey from classical physics to quantum gravity and relational views of reality.

Sean Carroll — *Something Deeply Hidden* Carroll presents one interpretation of quantum mechanics and argues for its ontological implications. It is valuable partly because readers can examine the distinction between accepted formalism and a strongly defended interpretation.

How to Use This Reading List

This section is not intended as a formal academic bibliography or as an endorsement of every argument contained in every listed work. The books and papers differ greatly. Some are historical foundations. Some are technical references. Some defend established interpretations. Some challenge them. Some are philosophical. Some are speculative.

They should be read according to their status. A useful reading discipline is to keep asking:

- What was observed?
- What was mathematically derived?
- What was experimentally confirmed?
- What interpretation has been added?
- What remains disputed?
- What problem is the author trying to solve?
- What would show the argument to be incomplete?

Readers exploring the Chavanian ecosystem should apply the same questions. No framework should be protected by the loyalty of its readers. A theory grows through comparison, criticism, formal development, and evidence.

The purpose of further reading is therefore not to gather authorities behind a preferred conclusion. It is to widen the field of inquiry.

- The established works show how physics reached its present strength.
- The Chavanian works ask whether another layer of synthesis may be possible.

Between them lies the real task of the reader: to respect what science knows, to recognize what it does not yet know, and to remain intellectually available to what reality may reveal next.

Acknowledgments

This book emerged from years of teaching physics, answering students' questions, and repeatedly returning to the meaning hidden behind familiar formulas.

I am grateful to every student who asked what was actually happening instead of stopping at the numerical answer. Their curiosity kept the subject alive for me.

I thank my family for the patience, space, and support that make sustained writing and independent inquiry possible.

I also acknowledge the generations of scientists, mathematicians, philosophers, teachers, and experimenters whose work created the foundations from which these questions could be asked.

Finally, I thank the readers who approach science with both respect and openness—willing to trust what has been established while continuing to examine what remains unexplained.

This book belongs to that continuing inquiry.

About the Author

Sandeep J. Chavan is a teacher, engineer, independent researcher, and author whose work explores physics, human behaviour, education, consciousness, social systems, and the deeper structures through which reality becomes observable.

He is the Founder–Director of Gyrus Institute of Science in Newasa, Maharashtra, where he has taught higher-secondary physics and science for more than fifteen years. His teaching approach begins with the physical event before moving toward definitions, quantities, equations, and numerical problems. He encourages students not only to calculate correctly, but also to ask what the calculation represents.

Before entering education, he worked in engineering and industrial process planning, including several years with Siemens. His experience across engineering, manufacturing, quality systems, and teaching shaped his interest in the relationship between theoretical knowledge and real-world consequence.

Chavan is the originator of Universal Energy Dynamics, a developing foundational framework that investigates whether matter, fields, forces, consciousness, and physical manifestations may arise from a deeper structured energetic reality. Within this framework, he proposes Structured Vacuum Energy as a possible universal physical foundation. His independent research includes writings on ripple dynamics, quantum phenomena, cosmology, matter formation, consciousness, psychology, leadership, constitutional systems, and the Law of Consequences.

He has written extensively across science, philosophy, education, psychology, leadership, and social thought through the wider Chavanian ecosystem. His books seek to connect technical clarity with accessible language while carefully distinguishing established knowledge, interpretation, and developing theory.

At the centre of his work is a simple conviction: knowledge should not merely provide answers; it should improve our ability to see the question beneath them.

Disclaimer and Statement of Originality

Disclaimer

This book is intended for educational, philosophical, and conceptual discussion.

It does not replace formal instruction in physics, mathematics, engineering, cosmology, quantum theory, or any other scientific discipline. Readers seeking technical mastery should consult standard textbooks, peer-reviewed research, qualified teachers, and recognized academic sources.

The discussions of force, energy, fields, spacetime, quantum theory, thermodynamics, vacuum, matter, and physical quantities are written for a general audience. Some explanations are simplified to make foundational questions accessible. Such simplification should not be treated as a complete technical account.

The book distinguishes among established science, interpretation, analogy, conceptual extension, and speculation. Nevertheless, readers should evaluate all scientific claims independently and consult primary sources where greater precision is required.

Universal Energy Dynamics, Structured Vacuum Energy, manifestation science, and related formulations presented in this book are developing ideas proposed by the author. They are not presented as universally accepted or experimentally established replacements for modern physics.

UED should be understood as a developing research framework. Its scientific value depends on further mathematical formulation, comparison with established theories, testable predictions, experimental investigation, and openness to revision or rejection.

References to SVE do not imply that its existence has been conclusively demonstrated. The term identifies a proposed foundational model through which questions concerning matter, fields, force, vacuum, stability, and physical manifestation may be investigated.

Analogies used throughout the book—such as maps, waves, whirlpools, flames, landscapes, and musical scores—are explanatory aids. They are not intended as literal physical models unless explicitly stated.

The book does not provide engineering, medical, legal, financial, or safety advice. No practical decision involving machinery, structures, electrical systems, health, laboratory work, or physical risk should be based solely on the conceptual discussions presented here.

The author and publisher accept no responsibility for loss, damage, injury, or misunderstanding arising from the misuse, misinterpretation, or unauthorized technical application of the material.

Scientific knowledge develops continuously. Interpretations, terminology, experimental results, and theoretical frameworks may change after publication. Readers are encouraged to compare the book with current scientific literature.

Statement of Originality

This work is an original conceptual presentation by Sandeep Chavan.

Its central arrangement, arguments, interpretations, explanatory structure, and formulations—including the treatment of physical quantities as manifestation vocabulary and the presentation of UED and SVE—have been developed as part of the author's wider independent research and writing.

The book draws upon established scientific concepts, historical developments, and philosophical questions that belong to the shared body of human knowledge. References to scientists, theories, experiments, and recognized works are included for educational and contextual purposes.

No claim is made to ownership of established scientific facts, standard equations, historical discoveries, commonly accepted terminology, or ideas already present in the scientific and philosophical literature.

The originality of this work lies in the author's synthesis, interpretation, organization, language, conceptual relationships, and proposed framework.

Any resemblance to other works in general themes—such as the limits of scientific models, the philosophy of measurement, emergence, vacuum structure, field ontology, or the distinction between mathematics and reality—may arise from engagement with common foundational questions.

The author has made a good-faith effort to distinguish established knowledge from original interpretation and developing theory.

Where the book presents UED or SVE, these ideas should be evaluated on the basis of clarity, evidence, mathematical development, predictive value, and scientific testability—not on the basis of authorship or authority.

This statement is not intended to prevent criticism, comparison, academic discussion, quotation, review, or independent investigation. Intellectual engagement is welcomed, provided the work is represented accurately and acknowledged appropriately.

All rights in the original expression, structure, and presentation of this book remain with the author, subject to applicable copyright law.

Don't miss out!

Visit the website below and you can sign up to receive emails whenever Sandeep Chavan publishes a new book. There's no charge and no obligation.

https://books2read.com/r/B-A-EPGPC-RQLLJ

BOOKS 2 READ

Connecting independent readers to independent writers.

Did you love *Physics: More Than You Know*? Then you should read *Universal Energy Dynamics*[1] by Sandeep Chavan!

[2]

Universal Energy Dynamics: A Consequence-Based Framework for Reality

Why do systems fail even when intelligence, technology, and planning are present?Why do crises appear suddenly after long periods of apparent stability?Why do actions produce consequences far removed from their origin?

In *Universal Energy Dynamics*, independent researcher and author **Sandeep J. Chavan** introduces a powerful interpretive framework for understanding how complex systems behave across physics, human institutions, and global networks.

---

1. https://books2read.com/u/mdBOnw
2. https://books2read.com/u/mdBOnw

Rather than explaining events through isolated causes or intentions, the book proposes a deeper structural model based on consequence-driven dynamics. According to this framework, systems evolve through the interaction of constraints, accumulated pressure, gradients of change, and threshold-driven correction.

At the center of Universal Energy Dynamics (UED) are several foundational concepts:

• **Delta** — unresolved pressure accumulating within a system• **Gradient** — directional differences that drive resolution• **Constraint** — limits that shape how systems evolve• **ATT (Alignment Threshold Time)** — the tolerance window before abrupt correction• **Consequence** — the structural feedback that restores equilibrium

Together, these principles provide a diagnostic lens for analyzing events that often appear chaotic or unpredictable.

Through conceptual explanations, diagrams, and real-world case studies—including financial crises, organizational failures, pandemics, climate systems, and information networks—**Sandeep J. Chavan** demonstrates how the same underlying dynamics operate across vastly different domains.

The book does not reject existing scientific theories. Instead, it places them within a broader structural perspective, showing how classical physics, relativity, quantum mechanics, and thermodynamics describe outcomes while Universal Energy Dynamics explores the deeper mechanisms that drive them.

By shifting focus from intention to consequence, this framework reveals why:

• stability can hide accumulating instability• systems fail after long periods of success• tightly connected networks amplify pressure• delayed feedback creates the illusion of randomness

Written in clear, disciplined language, *Universal Energy Dynamics* bridges the gap between science, philosophy, and systems thinking. It offers readers a new way to interpret events—from organizational decisions to planetary-scale challenges—through the lens of structural dynamics.

This book is not a manifesto or a prediction engine. It is a tool for understanding how systems accumulate pressure, cross thresholds, and correct themselves through consequence.

For readers interested in **systems science, philosophy of physics, complexity theory, and interdisciplinary thinking**, *Universal Energy Dynamics* provides a compelling framework for seeing reality in a new way.

Read more at https://www.sandeepjchavan.com/.

www.ingramcontent.com/pod-product-compliance
Lightning Source LLC
LaVergne TN
LVHW100726061126
851857LV00011B/409

* 9 7 8 8 1 6 8 9 7 7 6 3 1 *

www.ingramcontent.com/pod-product-compliance
Lightning Source LLC
LaVergne TN
LVHW100724061126
851857LV00011B/379